The Scientific Marx

The
Scientific
Marx

Daniel Little

University of Minnesota Press *Minneapolis*

The University of Minnesota Press
gratefully acknowledges assistance provided
by the Andrew W. Mellon Foundation
for publication of this book.

Published by the University of Minnesota Press
2037 University Avenue Southeast, Minneapolis, MN 55414.
Published simultaneously in Canada
by Fitzhenry & Whiteside Limited, Markham.

Library of Congress Cataloging-in-Publication Data

Little, Daniel.
  The scientific Marx.
  Bibliography: p.
  Includes index.
  1. Marx, Karl, 1818–1883. 2. Marx, Karl, 1818–1883.
Kapital. I. Title.
HX39.5.L56 1986   335.4   86-1384
ISBN 0-8166-1504-7
ISBN 0-8166-1505-5 (pbk.)

To my parents,
William and Emma Little,
and to the memory of
Abner Friedland

Contents

Acknowledgments

Thanks are due to colleagues and friends whose suggestions have greatly improved this book. Detailed criticisms and comments received from John McMurtry and James Farr on the penultimate draft of the manuscript were invaluable. I am also grateful for comments provided by Jerry Balmuth, Roland Blum, George Brenkert, Josh Cohen, Daniel Hausman, Colin Lawson, Georges Rey, and Huntington Terrell on various portions of the manuscript. I owe warm thanks to my wife Ronnie Friedland, both for her supportiveness and her editorial good sense. Parts of Chapters 1, 2, 4, and 7 have appeared in *The Social Science Journal* (vol. 22, 1985), *Topoi* (vol. 4, 1986), *Philosophy of the Social Sciences* (vol. 17, 1987), *The Southern Journal of Philosophy* (vol. 20, 1982), *Philosophy and Phenomenological Research* (vol. 42, 1981), and *Review of Radical Political Economics* (vol. 17, 1985-86). I gratefully acknowledge support provided by the Colgate University Junior Faculty Leave Program and the Colgate University Research Council. Meticulous copyediting was provided by Mary Byers.

Bibliographical Note

All references to Marx's writings will be included within the text, using the following abbreviations. Wherever possible I will refer to the Vintage Marx Library edition of Marx's writings (published in cooperation with Penguin Books and New Left Books).

Capital I: Karl Marx. *Capital: A Critique of Political Economy*, vol. I. Translated by Ben Fowkes. New York: Vintage, 1977.

Capital II: Karl Marx. *Capital: A Critique of Political Economy*, vol. II. Edited by Frederick Engels. Translated by David Fernbach. New York: Vintage, 1981.

Capital III: Karl Marx. *Capital: A Critique of Political Economy*, vol. III. Edited by Frederick Engels. Translated by David Fernbach. New York: Vintage, 1981.

Correspondence: Karl Marx and Frederick Engels. *Selected Correspondence*, 3rd ed. Edited by S. W. Ryazanskaya. Translated by I. Lasker. Moscow: Progress Publishers, 1975.

EW: Karl Marx. *Early Writings*. Edited by Quintin Hoare. Translated by Rodney Livingstone and Gregor Benton. New York: Vintage, 1975.

FI: Karl Marx. *The First International & After: Political Writings*, vol. III. Edited by David Fernbach. Translated by Paul Jackson, Joris de Bres, David Fernbach, Rosemary Sheed, and Geoffrey Nowell-Smith. New York: Vintage, 1974.

GI: Karl Marx and Frederick Engels. *The German Ideology*, part one, with selections from parts two and three. Edited by C. J. Arthur. Translated by International Publishers. New York: International Publishers, 1970.

Grundrisse: Karl Marx. *Grundrisse: Introduction to the Critique of Political Economy*. Translated by Martin Nicolaus. New York: Vintage, 1973.

HF: Karl Marx and Frederick Engels. *The Holy Family, or Critique of Critical Criticism.* Moscow: Progress Publishers, 1975.

PP: Karl Marx. *The Poverty of Philosophy.* Translated by International Publishers. New York: International Publishers, 1963.

R1848: Karl Marx. *The Revolutions of 1848: Political Writings*, vol. I. Edited by David Fernbach. Translation first published in 1888. New York: Vintage, 1974.

SE: Karl Marx. *Surveys from Exile: Political Writings*, vol. II. Edited by David Fernbach. Translated by Ben Fowkes and Paul Jackson. New York: Vintage, 1974.

TM: Karl Marx. *Texts on Method.* Translated and edited by Terrell Carver. New York: Harper & Row, 1975.

TSV I: Karl Marx. *Theories of Surplus Value*, vol. I. Moscow: Progress Publishers, 1968.

TSV II: Karl Marx. *Theories of Surplus Value*, vol. II. Moscow: Progress Publishers, 1968.

TSV III: Karl Marx. *Theories of Surplus Value*, vol. III. Moscow: Progress Publishers, 1968.

WLC: Karl Marx. *Wage Labour and Capital & Value, Price and Profit.* New York: International Publishers, 1972.

The Scientific Marx

Introduction

Marx presented *Capital* to the public as a work of science: a controlled, rigorous investigation of the "inner physiology" of the capitalist economic system. What distinguished *Capital* from the writings of other contemporary socialists, he believed, was the empirical precision and objectivity of his treatment. This claim of scientific standing has two major parts. First, Marx believed that he had provided the basis for a scientific explanation of crucial aspects of the capitalist economy. Using his analysis of the central mechanisms of the capitalist system, Marx attempted to predict and explain some of the most important patterns of capitalist development – the law of the falling rate of profit, the creation of an industrial reserve army, the polarization of classes, the concentration and centralization of property, and the increasing severity of cyclical crises. Second, Marx intended that his work be evaluated by ordinary standards of scientific adequacy, and he believed that the available empirical evidence would in fact vindicate his conclusions. His theory of ideology notwithstanding, Marx thought that scientific objectivity was possible and that scientific research could offset the biases associated with class position. And the objectivity of his own analysis, he believed, derived from the rigor of his reasoning and the empirical evidence that could be mustered in favor of his account.

In the late twentieth century many deficiencies in Marx's scientific claims seem to loom large. Many of the predictions offered in *Capital* have failed to materialize within late capitalism: The rate of profit is more stable than Marx expected; classes are less sharply defined and more harmonious; labor unions have acquired substantial power against the owners of capital; the capitalist state seems more capable of managing the economic crises of capitalism than Marx expected; and socialism seems to be beyond the horizon for the Western capitalist nations. These failures of prediction lead many to believe that Marx's claim to having advanced a scientific theory of the capitalist system has been refuted.[1] On the theoretical side equally severe challenges have been put forward against

Marx's scientific work. Since Hilferding and Böhm-Bawerk economists have criticized the labor theory of value; Karl Popper has argued that Marx's account of capitalism is unscientific because it is unfalsifiable; various authors have accused Marx of a form of economic reductionism; and some Marxists (e.g., Marcuse and Wellmer) have derided Marx's aspiration to scientific status as a form of "scientism." Few Marxists have continued to emphasize Marx's scientific claims; most have preferred to stress his humanism or his theory of historical materialism.

This book thus represents a break with many current views of Marx's analysis of capitalism in that it takes seriously his claim that *Capital* is a rigorous scientific analysis of the capitalist mode of production. I regard *Capital* as a classic and substantial work of social science, and I accept Marx's own view that this analysis plays a crucial role in his system as a whole. In what follows I will discuss the main features of Marx's account applying the tools of contemporary philosophy of science. Like any scientific effort, Marx's work implicitly defines a theory of scientific knowledge through its conceptions of subject matter, explanation, and justification. My aim is to reconstruct the details of this theory. What range of phenomena is Marx's analysis intended to explain? What are the purposes and scope of Marx's research? How is Marx's account organized? Is it a unified deductive theory? What sorts of laws does it rely on? What kinds of explanations does it present? How is it empirically justified? And how does *Capital* relate to the larger concerns of historical materialism?

A careful look at *Capital* as a work of social science is important for several reasons. First, even a cursory reading of Marx's writings makes it clear that his claim to having conducted a scientific analysis of the capitalist mode of production is central to his system as a whole. His theory of historical materialism requires that empirical investigation replace social philosophy as a basis for social criticism; his critique of the utopian socialists turns on their lack of a scientific analysis of existing institutions; and his theory of revolution requires that the party of the proletariat base its strategies on a scientific understanding of the capitalist order it seeks to overthrow. Finally, his belief in the inevitable demise of capitalism depends on his discovery in *Capital* of the sources of ever-deepening crisis to be expected within the capitalist economy. In each case the claim that *Capital* represents an objective empirical study of capitalism plays a central role. If this contention is unjustified, Marx's case is greatly weakened. So it is crucial to come to a detailed and accurate view of the scientific characteristics of *Capital* in order to assess Marx's system as a whole.

Second, a detailed analysis of Marx's scientific work has implications for the philosophy of social science more generally. Recent philosophers of science have held that careful study of particular cases in the history of science is a fruitful way of coming to new insights into the nature of scientific reasoning. This sort of study is especially useful as a corrective for the "unity-of-science" tendency implicit in much work in the philosophy of social science, that is, the assumption that the logic and organization of social science should resemble that of the natural sciences. The problems we will encounter in *Capital* concerning its standing as a work of social science are to some degree representative of other social sciences as well. Thus some of the findings presented here may help illuminate important problems in the philosophy of social science more generally. For example, the reading of Marx's work provided here will suggest that the concept of "theory" appropriate to natural science is less suitable for social science; that the logic of justification is different in social science and in natural science in that the hypothetico-deductive model is inappropriate in the former; that explanations in social science often depend on a "logic of institutions" rather than laws of nature; that the notion of a "hypothetical construct" is less useful in social science than in natural science; and that the organization of knowledge in social science is significantly different from that in natural science.

Although the scientific standing of *Capital* is fundamental to an understanding of Marx's writings, this issue has not been extensively discussed. Few philosophers of science have seriously considered Marx's work as a body of scientific knowledge. (The most prominent exception is Karl Popper, whose views will be discussed in later chapters.) Perhaps more surprisingly, few interpreters of Marx have considered the scientific standing of his work in satisfactory detail, preferring instead to emphasize his theories of alienation or historical materialism, or technical issues in his economic theory.[2]

Fortunately, recent developments in Anglo-American Marx scholarship have made conditions ripe for a study of Marx's science grounded in analytic philosophy of science. A great deal of valuable work in Marx studies in the past ten years has originated in an analytic approach to Marx's basic ideas. Full-scale treatments of his work by analytic philosophers (e.g., G. A. Cohen, Allen Wood, George Brenkert, and Allen Buchanan)[3] have raised the level of rigor expected in discussions of Marx. Several related trends are particularly important for the purposes of this book. First, a great volume of detailed work on the theory of historical materialism has appeared in recent years. Gerald Cohen, John McMurtry, William Shaw, and others[4] have provided exacting treatments of the central ideas of historical materialism and the sorts of explanations offered in this part of Marx's thought. This literature has greatly clarified many of the central concepts and theses of classical historical materialism, and it has raised old controversies in a greatly sharpened form (e.g., disputes over the relative primacy of the forces and relations of production in historical change, and disputes over the logical character of the explanations contained in historical materialism—functional, causal, teleological).[5]

These writings bring us to the brink of Marx's science—but only to the brink. For historical materialism directs Marx to provide detailed studies of the economic structures of various modes of production; but it does not provide guidance for conducting such an investigation, nor does it contain the concepts and theoretical tools necessary to formulate clear explanatory hypotheses in this area. The existing literature on historical materialism does not consider these questions at great length; rather, authors tend to treat *Capital* as an addendum to historical materialism. However, *Capital* goes substantially beyond Marx's theory of historical materialism, both methodologically and substantively. In his economics Marx develops a rigorous conception of the sort of explanatory system needed for a scientific understanding of the capitalist economy; and he provides a markedly more precise and developed treatment of the economic structure of capitalism than is contained in the chief writings of historical materialism. This body of work requires separate treatment; it cannot be simply subsumed under our current understanding of the theory of historical materialism.

A second important area of research within recent Anglo-American Marx studies may be loosely described as a "microfoundational" approach to Marxian historical and social explanations. This body of research has developed in close proximity to the aforementioned discussions of historical materialism (indeed, a central issue between the two groups is the utility of functional explanations in Marxism). Noteworthy contributions in this area include John Roemer's work on the theory of exploitation;[6] Jon Elster's efforts to apply the techniques of rational choice theory and game theory to Marxian explanations;[7] Robert Brenner's microclass analysis of the emergence of capitalist property relations out of feudal property relations;[8] and Erik Olin Wright's work on the relations between class and politics.[9] These works are highly relevant to this book. Indeed, one of the central views to be argued here is that Marx's explanations take the form of analyses of a "logic of institutions" that flows from the circumstances in which individuals of different classes find themselves (chapters 1 and 5); and this view finds much support in the work of these authors. What these authors do not seek to do, however, is to arrive at the logic of explanation, justification, and method of Marx's scientific work as a whole. Consequently their work complements but does not supplant the central concerns of this book.[10]

Finally, there continues to be a dynamic literature on Marxian economics. Moreover, this literature has been particularly robust in recent years as a result of the efforts of Michio Morishima, Ian Steedman, Robert Paul Wolff, John Roemer, and others.[11] Here, however, the emphasis is generally on the technical particulars of Marx's economic analysis, not the larger claim that the whole constitutes a scientific theory of the capitalist economic system. Thus Marxian economists consider the strengths and weaknesses of the labor theory of value; they consider Marx's achievements as a mathematical economist in his "schemes of reproduction"; they analyze his theory of crisis; they consider the adequacy of various solutions to the transformation problem; and so forth. But what is lacking in such accounts is any discussion of questions having to do with this work as an organized body of scientific knowledge. Unlike the Marxian economists, I will consider Marx's theory of capitalism as an integrated whole, not merely distinct pieces of economic analysis. (In fact, I will hold that the core of Marx's account has to do with the social institutions defining the capitalist mode of production, not the quantitative economic conclusions he reaches.)

It should be noted that European Marxists have paid more attention to the logic of *Capital* than have Anglo-American scholars. Particularly prominent are the writings of Louis Althusser and others influenced by him—Hindess and Hirst, Poulantzas, Godelier, to name a few.[12] Unfortunately these works are often too burdened by miscellaneous philosophical theories to be of great value in understanding Marx as a scientist. These writers have tended to emphasize exotic epistemological or methodological ideas in explicating Marx's views rather than the historical and empirical method he pursues. As a result, the Marx described in *Reading Capital* is more philosophical and less plausible than the Marx found in *Capital*. Against these commentators, I will hold that Marx's work does not depend on esoteric views about philosophy or methodology; rather, his account is a straightforward attempt to make sense of the empirical phenomena of capitalism.

Some comment is needed concerning the method of investigation to be used throughout this book. My ultimate purpose is to discover the particulars of Marx's theory of science: his conceptions of scientific method and inquiry, of justification and explanation, and the like. Throughout Marx's works are sprinkled specific discussions of many of these topics.[13] One might suppose, therefore, that an appropriate beginning for my investigation would be to work out in detail Marx's explicit theory of science, insofar as such a theory can be reconstructed from his overt comments about method in *Capital* and elsewhere. At various points I will indeed address some of Marx's explicit methodological ideas (particularly in chapter 4, "Essentialism, Abstraction, and Dialectics"). But in general I doubt that these methodological comments give a reliable view of the theory of science that actually guided Marx's research. In analyzing the conception of science represented by any body of scientific work, we need to distinguish between the scientist's explicit theory of science and the implicit theory that underlies the scientific practice. The former is to be found, for example, in the scientist's statements of method and his criticisms of other scientists. The latter, by contrast, is embodied in the investigator's actual scientific practice. This implicit theory is constituted by the set of standards and assumptions that actually guide the conduct of the research and theory formation, and it is largely independent of the explicit methodological ideas advanced by the investigator.

As Thomas Kuhn shows, the implicit theories of science underlying a given corpus of scientific research—the "paradigms"—are complex and obscure, and it is a major problem of research and interpretation for the historian of science to reconstruct the details of these theories.[14] Moreover, such paradigms operate at an unconscious level (much in the way that native speakers' knowledge of the grammar of their language is unconscious).[15] Consequently there is no reason to expect that practitioners of science are in a privileged position to be able to articulate the guiding assumptions about science that give structure to their work. And in fact the historian of science often finds that gifted scientists may have views about the logical character of their own research that differ widely from their actual practice.[16] It therefore would be surprising if Marx's methodological comments succeeded in elaborating the complex set of rules and methods guiding his empirical and theoretical practice.

In light of these considerations, my investigation will be based largely on a different sort of evidence: evidence taken from Marx's actual practice as a social scientist rather than from his methodological writings. *Capital* constitutes a rich body of explanations, justificatory arguments, inferences, descriptions, economic analyses, and the like; and these "data" implicitly define a structured theory of scientific reasoning. My approach will involve, first, careful consideration of these sort of data, and second, an attempt to reconstruct the assumptions these data represent about the nature of science. Thus when I consider the logic of explanation in *Capital*, I will turn, not to Marx's explicit comments about explanation throughout his work, but to particular explanations in *Capital*. These concrete examples reveal a very definite logic, and it is a pattern of explanation which Marx almost never formulates in any of his comments about method.

These qualifications should not be taken to suggest that Marx's methodological writings are generally at odds with his theory of science; on the contrary, Marx is generally a good guide to scientific method and his own practice. But wherever there is a gap between the explicit and implicit theories of science, it is the implicit theory that is of primary importance. The point is therefore one of focus. My concern is with the theory of science embodied in Marx's practice as a working social scientist, and the best evidence for determining the particulars of that conception is found in his practice, not in his methodological pronouncements.

For similar reasons I will not spend a great deal of time considering Marx's methodological or philosophical criticisms of some of his competitors (e.g., J. S. Mill, Auguste Comte, and Proudhon). (Once again, chapter 4 represents the chief exception; there I will consider Marx's critique of orthodox political economy.) Marx's criticisms of his competitors are important in a variety of contexts – for example, as a way of determining his background philosophical ideas and assumptions. However, they cannot be taken to define the theory of science implicit in Marx's practice in *Capital*. In order to do that, one would need to establish that the explicit theory attributable to these criticisms is identical to the implicit theory of science in *Capital*; but in many important details

I believe they are not identical. It is the *implicit* theory of science supporting *Capital* that is of central concern here, and from this point of view, Marx's express methodological comments and his philosophical criticisms of his competitors are of secondary importance. The primary evidence on which to construct an interpretation of Marx's theory of science must be the body of concrete scientific reasoning and explanation to be found in *Capital*.

We can summarize the issues raised in this book under three headings. First, how does Marx define the aims of his investigation? What does he seek to explain? What limits does he impose on the scope of inquiry? What is the logical form of his account? And how does *Capital* relate to the full theory of historical materialism? Chapters 1 and 2 concern themselves with this family of issues. Second, what is Marx's explanatory paradigm in *Capital*? What constitutes an explanation for Marx? On what sorts of regularities do his explanations typically depend? Chapters 3, 4, and 5 are directly concerned with these issues. And third, what is Marx's conception of empirical justification? Through what sorts of arguments does he endeavor to establish the correctness of his analysis of capitalism? Chapter 6 describes Marx's use of evidence in evaluating his construction, and chapter 7 considers criticisms offered by Karl Popper and E. P. Thompson against Marx's empirical practices. Having discussed Marx's positions on these issues, we will have described the chief elements of Marx's theory of science as well; for these elements—the intentions that define his plan of research and the assumptions about the logical structure of theory, explanation, and justification that guide his research—largely define a conception of scientific inquiry and knowledge.

My general conclusion will be that *Capital* is a careful, rigorous, and developed work of empirical social science that goes substantially beyond Marx's writings on historical materialism in both rigor and detail of analysis. *Capital* is guided by a sober definition of research goals; it advances a specific and developed account of the capitalist economic structure; it offers substantive explanations of some of the phenomena of the capitalist economy; and it makes appropriate use of empirical evidence in evaluating the account. Further, I will argue that many common objections against the scientific standing of Marx's analysis are invalid. In particular, I will rebut the charges that Marx's research depends on suspect methodological ideas (e.g., Hegelian dialectical reasoning), and I will argue that failed predictions in general do not falsify Marx's account (because of general problems of using predictions to evaluate theories in the social sciences).

Marx's *Capital* is thus a vigorous contribution to our knowledge of capitalism and a noteworthy achievement in social science. That being said, I must forewarn that a theme throughout this work is that of the limits of Marx's purposes and achievements. Among my contentions will be that Marx's account of capitalism is not a "theory" in the sense found in natural science and that its long-term predictions are less firmly attached to his basic analysis than is sometimes thought; that Marx's account is confined to the economic structure of capitalism and has only rather limited implications for capitalist society as a whole; that Marx's economics must be sharply distinguished from the broader claims of historical materialism; and that Marx overestimates the importance and fruitfulness of the labor theory of value. It is only by recognizing the limitations on the scope of Marx's views that we may properly appreciate his achievements. By overstating the power of Marx's theories, many of his defenders have led his critics to underestimate these theories' power and worth. My intention here is to provide a balanced interpretation of Marx's achievements as a social scientist. If Marx is to be of continuing importance for social science, this can only be within the context of a clear and critical appraisal of his strengths and weaknesses.

Further, though I have singled out Marx's claim to have provided a scientific treatment of capitalism for special attention, it should be noted that I do not maintain that this is the only "authentic" Marx or that other aspects of the system do not require careful attention. My sole assumption is that this portion of Marx's program is centrally important, that it has been inadequately studied in the literature, and that it can be usefully considered in isolation from other parts of Marx's system.

1

Naturalism and *Capital*

It is often assumed that the scientific standing of *Capital* rises and falls according to its standing as a *theory* in the sense appropriate to natural science. This view of Marx's work can be described as a form of *naturalism*. According to this sort of account, *Capital* is intended to provide a unified deductive theory of capitalism; this theory formulates an organized set of hypotheses about the underlying mechanisms of the capitalist mode of production; the theory permits the derivation of specific long-range predictions about the development of the capitalist system; and the theory is confirmed through the truth of its consequences for experience.

This view of *Capital* is almost ubiquitous among commentators who have written on Marx's science. Thus Lenin writes, "Historical materialism made it possible for the first time to study with the accuracy of the natural sciences the social conditions of the life of the masses and the changes in these conditions."[1] Marx's theory of capitalism formulates the "objective laws governing the development of the system of social relations," and permits us to "deduce the inevitability of the transformation of capitalist society into socialist society wholly and exclusively from the economic law of motion of contemporary society."[2] Karl Popper's treatment of Marx's system as "historicist" sounds a similar theme; he too judges that Marx intended to predict the behavior of capitalism in much the way that the science of mechanics foretells states of mechanical systems. "Marx's 'inexorable laws' of nature and of historical development show clearly the influence of the Laplacean atmosphere."[3]

Lest it be supposed that this approach to Marx is confined to early, unsophisticated treatments, however, consider the recent views of Russell Keat and John Urry: "[Marx's purpose is] the elucidation of the internal structure of each mode of production, a theoretical activity involving the positing of models of the relevant causal structures and mechanisms." [4] Thus Keat and Urry regard Marx's as a highly theoretical project: to formulate a set of hypotheses about the real but unobservable structures that define the capitalist mode of production, and to work out the causal consequences of these structures. In a similar vein, Derek Sayer writes, "I have argued that Marx's notion of explanation is . . . a realist one. . . . My point is simply that since Marx's explanations do involve laws, they are capable of generating predictions, and if they can generate predictions, they can also be tested by them."[5] Sayer distances himself from positivist conceptions of theory by insisting that Marx's theories must be realistically interpreted; but he too endorses the idea that Marx's analysis of capitalism should be formulated as a unified deductive theory and that it explains the phenomena of capitalism by showing how those phenomena may be subsumed under the theoretical laws put forward in *Capital*.

What these interpretations share is a highly abstract (and I will argue, erroneous) conception of the logical form of the knowledge of capitalism offered in *Capital*. Each author presupposes that Marx advances an organized, abstract theory of capitalism, and then works out the empirical consequences of this theory for the observable phenomena of capitalism in the form of a set of predictions. The conception of knowledge at work here is broadly speaking *naturalistic* in that it conceives of Marx's analysis of capitalism in strong analogy with certain types of theories in natural science (especially classical mechanics). It construes Marx's account as a deductive system capable of unifying and explaining most capitalist phenomena on the basis of a small number of theoretical assumptions. It supposes that Marx identifies the unobservable mechanisms which determine the phenomena of capitalism, and that the account is explanatory because it advances a set of social laws that are analogous to laws of nature. Finally, it suggests that Marx's account may achieve the degree of precision and predictive power associated with natural science in foretelling the behavior of capitalism.[6]

This chapter rejects all the main premises of this interpretation of Marx's work. Marx's construction in *Capital* is not a general system based on a few hypotheses; it does not unify all capitalist phenomena within a deductive system; it does not identify a set of "laws of motion" that are genuinely analogous to laws of nature; and it makes use of empirical evidence in a way quite different from that appropriate to the hypothetico-deductive model of justification. These conclusions do not reduce the scientific standing of Marx's construction, however, because this form of naturalism is in general illsuited for social science. On a more adequate analysis of scientific knowledge (one that does not attempt to see all scientific knowledge on the model of the unified deductive theory) Marx's account may be viewed as perfectly appropriate: a selective description of certain fundamental features of the capitalist economy and an examination of the "logic of institutions" imposed by those features on the economy as a whole.

Nonnaturalistic Features of *Capital*

What Is Naturalism?

The issue of naturalism has been much discussed in the philosophy of social science, and it has had special importance in current Marx studies. In a recent work on naturalism David Thomas defines the problem of naturalism in the following way: "Can social study conform to a naturalistic methodology, that is replicate the methodology of natural science? Is naturalism, the doctrine that there can be a natural scientific study of society, correct?"[7] Thomas argues for an affirmative answer to this question. Roy Bhaskar defines the issue in very much the same terms as Thomas. "To what extent can society be studied in the same way as nature? . . . Naturalism may be defined as the thesis that there is (or can be) an essential unity of method between the natural sciences."[8]

On these accounts naturalism amounts to the view that the methods of the social sciences are (or may be, or must be) the same as those of the natural sciences. But in what respects are the methods of the social and natural sciences thought to be similar? Plainly there will be differences between the methods of inquiry appropriate to the two areas of science; for example, the social sciences usually cannot make use of the methods of repeatable controlled experimentation that are characteristic of physics.[9] (Differences of these sorts are not unique to the contrast between social and natural science, however; for example, meteorology and high-energy physics differ in this way as well.) Rather, the relevant similarity in method presumably has to do with the general logic of science instead of the particular methods of inquiry. Keat and Urry put the point this way: "[In naturalism] there is an underlying claim that there is only one logic of science, to which any intellectual activity aspiring to the title of 'science' must conform."[10]

The "general logic of science" referred to here might include such abstract features of scientific knowledge as the logical form of organization of scientific knowledge, the relation between theory and evidence, the nature of scientific explanation, and the character of scientific concepts. This formulation leaves a good deal of latitude in specifying naturalism more exactly, however. Different theories of science postulate different "logics of science," and there will be different forms of naturalism corresponding to these different conceptions of the defining features of the natural sciences. Keat and Urry make this point in distinguishing between positivist and nonpositivist forms of naturalism.[11] I will not hold that every imaginable form of naturalism is false; indeed, I expect that the "logic of science" may be specified in a sense innocuous enough to encompass all scientific knowledge—social and natural. Instead, I will single out several features of common forms of naturalism that are particularly inappropriate in application to Marx's system.

The account I will oppose might be called predictive-theory naturalism (P-T naturalism). On this view, scientific knowledge typically takes the form of deductive theoretical systems and associated bodies of empirical consequences. This form of naturalism can be summarized through the following theses.

1. Scientific knowledge takes the form of organized theories which possess a unified deductive structure.
2. Such theories typically describe unobservable mechanisms in order to explain observable conditions.
3. Such theories attempt to formulate laws of nature.
4. These theories make relatively precise predictions.
5. These theories achieve empirical corroboration through their deductive consequences for experience.[12]

This conception of the logic of science has several important sources in the history and philosophy of science. Most evidently it is closely related to the enlightened empiricist philosophy of science of late logical positivism.[13] But though logical positivism has developed this theory in great detail, P-T naturalism does not originate in logical positivism; rather, it represents a relatively accurate view of the nature of the physical sciences since Galileo and Newton. Moreover, a reading of postpositivist philosophy of science shows that many aspects of P-T naturalism continue to characterize avowedly antipositivist theories of science.

Carl Hempel's formulation of the logical structure of a scientific theory is representative of enlightened positivist versions of P-T naturalism. He writes, "We will assume that theories are stated in the form of axiomatized systems as here described; i.e., by listing, first the primitive and the derivative terms and the definitions for the latter, second, the postulates. . . . The classical paradigms of deductive systems of this kind are the axiomatizations of various mathematical theories, such as Euclidean . . . geometry . . . ; but by now, a number of theories in empirical science have likewise been put into axiomatic form."[14] Ernest Nagel's description of classical mechanics reveals the same basic conception. "The science of mechanics was the first branch of mathematical physics to assume the form of a comprehensive theory. The success of that theory in explaining and bringing into systematic relation a large variety of phenomena was for a long time unprecedented; and the belief entertained by many eminent scientists and philosophers . . . that all the processes of nature would eventually fall within the scope of its principles was repeatedly confirmed by the absorption of several sectors of physics into mechanics."[15] On this account the content of a scientific theory is encapsulated in the (it is hoped small) set of axioms or theoretical principles; these principles are then worked out deductively in order to establish their consequences for experience. And the central task of science is to systematize the bewildering array of empirical phenomena under a few general laws. "A theoretical explanation . . . reveals the different regularities exhibited by a variety of phenomena."[16]

A second feature of P-T naturalism is the use of theoretical concepts and unobservable entities. Explanation in the natural sciences typically is based on a theory of a set of unobservable mechanisms or entities whose properties give rise to the observable phenomena. Hempel represents a common view when he writes, "It is a remarkable fact . . . that the greatest advances in scientific systematization have not been accomplished by means of laws referring explicitly to observables, . . . but rather by means of laws that speak of various hypothetical, or theoretical, entities."[17] The great successes of natural science (the theory of the atom, quantum mechanics, genetic theory, and so on) all have depended on the postulation of hypothetical entities or processes as the basis for explanation of observable phenomena.

Finally, P-T naturalism postulates a general theory of confirmation for scientific theories. Since scientific theories consist of hypotheses that cannot be directly confirmed or tested, P-T naturalism holds that scientific theories are to be empirically evaluated as wholes. The theory is a unified deductive system with consequences for experience; thus P-T naturalism postulates that the theory is to be evaluated in terms of the relative degree of success of its empirical predictions. More fully, the theory consists of a set of general laws. When conjoined with appropriate bridge laws and boundary-condition statements, these laws entail a class of observational and experimental predictions. And the theory as a whole is corroborated according to the degree of success found in its family of predictions. Here again Hempel's writings are fairly representative of the theory of confirmation contained in the late positivist theory of science.[18]

Lest it be supposed that P-T naturalism is specific to enlightened positivism, two points should be noted. First, the account put forward by enlightened positivism has a great deal of historical authenticity. P-T naturalism is a relatively accurate description of many of the most important achievements of modern natural science. Since Newton the physical sciences have been highly theoretical and abstract; they often have possessed just the sort of deductive unity and rigor attributed to them by P-T naturalism; they have postulated unobservable entities in order to explain observable phenomena (atoms, forces, genes); they have permitted highly precise predictions (in astronomy, mechanics, atomic physics, etc.); and they have been justified perforce only through their observational consequences. Thus the history of the natural sciences since Galileo and Newton has been just the sort of search for deductive systematization that Nagel and Hempel describe.[19] So this aspect of P-T naturalism can (with appropriate qualifications) be taken to be a fair description of important features of the general program of the natural sciences in the past three centuries.

Second, however, most postpositivist philosophy of science has tacitly accepted much of this deductivist account as correctly characterizing the logic of science. Several large families of criticisms have been offered against logical positivism, but these objections have commonly presupposed the correctness of the main features of P-T naturalism. Postpositivist philosophy of science has tended toward realism in the interpretation of theoretical terms; it has offered damaging criticism of the positivist distinction between observation and theory; and it has tended to be more sensitive to the history and sociology of science.[20] But it has not rejected the basic idea that scientific knowledge takes the form of unified deductive systems or that scientific theories formulate laws of nature or that theories are confirmed chiefly through their observational consequences. Thus postpositivist philosophy of science differs in fundamental respects from positivist philosophy, but it largely accepts the assumptions identified earlier as P-T naturalism.

It is especially important for our purposes to note that P-T naturalism is broad enough to include important features of the positions of many current antipositivist interpreters of Marx. In particular, the realistic interpretations of Marx's science now in wide circulation share many of the features of this form of naturalism. Keat and Urry represent a developed version of the realist interpretation of Marx's system, and similar views can be found in the writings of Roy Bhaskar and Derek Sayer.[21] These are among the most extensive and important contemporary treatments of Marx's social science; and they are dependent on many of the assumptions identified here as naturalistic. These views will be discussed later in this chapter.

When applied to *Capital*, P-T naturalism leads to the expectations that Marx's analysis of capitalism must be a unified deductive theory; that this theory is intended to unify and explain most capitalist phenomena on the basis of a small number of theoretical assumptions; and that this theory seeks to identify the unobservable mechanisms that determine the phenomena of capitalism. The account is explanatory because it lays bare a set of social laws analogous to laws of nature. Finally, P-T naturalism suggests that Marx's account can achieve the degree of precision associated with natural science in predicting the behavior of capitalism, and that this theory is properly confirmed or falsified through the success or failure of its predictive consequences. In the following, however, I will argue that this interpretation of Marx's system is mistaken at virtually every step.

No Unified Deductive Theory

According to the previously described forms of naturalism, it should be possible to represent *Capital* as a unified deductive theory in which a small number of theoretical principles serve to unify the empirical phenomena of capitalist society. Even a cursory survey of *Capital* shows that Marx's system does not possess a unified deductive structure of this sort, however. Interpretations that strive to represent *Capital* in this manner are compelled to discard much of the material of *Capital* as unnecessary or merely illustrative.

One example of this sort of attempt to unify Marx's system as a single comprehensive theory is the school of thought that construes Marx as a pure political economist. Typically such interpretations identify the theoretical core of Marx's economic theory (generally the labor theory of value) and then attempt to construe the rest of *Capital* either as the deductive articulation of these theoretical principles or as the attempt to resolve empirical problems making use of these principles.[22] A second important example of this "theoretist" approach to *Capital* can be found in structuralist Marxism, particularly that of Althusser and his followers. In this case, instead of an economic interpretation of Marx's system, we find an effort to describe *Capital* as a general theory of the "structures" that define and animate the capitalist mode of production. For example, Hindess and Hirst hold that *Capital* is fundamentally an abstract theory of the capitalist mode of production that derives the "logic" of the system from the concept of the mode of production. Here too the aim is to portray *Capital* as a unified set of theoretical principles, with the rest of the work being treated as illustrative material or derived consequences.[23] This account shows the same predisposition identified earlier to construe *Capital* as an organized theoretical system, and the same reductionist necessity to downplay those portions of the work which cannot be easily assimilated to the theoretical model.

On my view, there is no single set of theoretical principles under which the bulk of *Capital* can be subsumed. Consequently no unified-theory interpretation of *Capital* is correct. Rather, *Capital* consists of a large number of relatively independent observations, descriptions, analyses, and the like, of different aspects of the capitalist economy. There is an irreducible plurality of the factors described, and there is no intention of reducing the whole of the capitalist system to a few theoretical premises.

This point emerges most clearly from a simple inventory of the wide variety of different sorts of material found in *Capital*. Consider for example the diverse concerns contained in part one of *Capital*.[24] Part one opens with Marx's treatment of the commodity form, his development of the labor theory of value, and his distinction between abstract and concrete labor. Marx then turns to the "fetishism of commodities." Here he attempts to account for the "opacity" of the value form (the circumstance that participants fail to recognize the nature of the social relations underlying a commodity-producing regime). The tone of technical economic analysis is largely absent here, and Marx's account of fetishism does not depend on the particulars of the labor theory of value advanced in the preceding pages. Rather, he explains fetishism in terms of the social relations supporting commodity production. "This fetishism of the world of commodities arises from the peculiar social character of the labour which produces them" (p. 165). In following chapters Marx turns to the process of circulation and exchange within a commodity-producing economy. This discussion includes material on the history of money (pp. 221–40), the function of law within a commodity system (p. 178), the function of money and credit in the world economy (pp. 238–44), and the abstract logic of commodity circulation (pp. 198–220).

The diversity to be found in part one of *Capital* is reproduced throughout much of the remainder of the work as well. An examination of the central themes shows that Marx's construction includes at least the following elements:

1. A description of the property system of capitalism: privated property in productive wealth and labor power.
2. A description of the purpose of production within capitalism: commodity production for profit and accumulation.
3. A developed treatment of the labor theory of value.
4. An abstract model of the capitalist mode of production (CMP) that embodies these features couched in terms of the labor theory of value.
5. A description of the workings of a competitive market.
6. An analysis of the economic and social implications of these features of the capitalist economy (crisis, the industrial reserve army, the falling rate of profit).
7. A sociological account of how the property relations of capitalism are reproduced.
8. A historical account of how these property relations were established within precapitalist society (primitive accumulation).
9. A description of the conditions of life and work of the working class (the working day, the effects of industry on the worker, etc.).

The variety displayed in these different elements of Marx's analysis in *Capital* shows that his treatment of capitalism does not take the form of a unified deductive system from which all relevant particulars can be deduced. Instead, Marx's account is a family of related explanatory arguments, bits of analysis, historical comments, and descriptive efforts loosely organized by a common perspective. No general theory akin to atomic theory permits Marx to unify all his material into a single deductive system. To be sure, there are connections between these various efforts. The economic analysis is intended to embody the social assumptions of Marx's account of the capitalist mode of production; the descriptive matter is thought to establish the correctness of these assumptions; the historical matter is intended to show how these structural conditions were established; and so forth. But these connections are relations among independently justified bodies of knowledge—not deductive relations of subsumption of empirical matter under general theoretical principles.

It is undeniable that Marx has various "theories" (a theory of value, a theory of exploitation, a theory of the wage, a theory of ground rent, a theory of capitalist motivation, a theory of unemployment, a theory of primitive accumulation, a theory of crisis). But none of these is *the* theory of *Capital*. Rather, each is a single hypothesis, or set of hypotheses, relating to particular and limited aspects of the domain of phenomena of the capitalist economy. In order to represent Marx's construction in *Capital* as a unified theory, one of two circumstances must obtain. Either it must be possible to identify a more fundamental set of assumptions or hypotheses to which these diverse explanations and analyses might be reduced; or else it must be possible to represent Marx's theory as the conjunction of all the logically independent assumptions and hypotheses advanced anywhere in *Capital*. The first alternative is entirely unpromising because the explanations, descriptions, and analyses identified here appear to be largely logically independent; that is, it is possible to hold or reject various of them independently from the others. It follows from this that there is no higher-level theory to which these more particular hypotheses and explanations might be reduced. These various explanations are not simply special cases of some more general and comprehensive theory; they are not derived consequences from a higher-level set of theoretical principles. In particular, it will be held in chapter 3 that most of Marx's explanations are logically independent of the labor theory of value (LTV, the most common candidate for a "general theory" in Marx's system) in that it is possible to reject the LTV while retaining these more specialized explanations.[25]

The second alternative is logically unobjectionable (though there is a serious possibility that the resulting system will be inconsistent). The result of this logical reconstruction, however, does not usefully illuminate the character of Marx's analysis of capitalism in *Capital*. This is so in several respects. First, the set of assumptions to be identified as basic theoretical premises would be extremely large, since there is no point in *Capital* at which Marx's theory is "finished," with further discussion taking the form of deriving consequences from a completed theoretical system. Rather, Marx continues to introduce assumptions throughout *Capital*. Therefore it would be as difficult to summarize the reconstructed theory as it is to summarize the whole of *Capital*; this reconstruction thus provides no useful theoretical economy. More important, however, this reconstruction badly misrepresents Marx's process of inquiry. The "unified-theory" model is illuminating in the natural sciences because in some important respect it reflects the process of investigation characteristic of physical science. Much research in the physical sciences takes the form of an experimental testing of the logical and mathematical consequences of a small set of theoretical assumptions. Marx's work in *Capital*, however, does not take this form. Marx is not typically deriving consequences from an established theoretical system; rather, he is offering a novel analysis of a feature of the capitalist economy that rests to some extent on previously established results and to some extent on new analytic insights.

Thus a survey of the main topics and explanations advanced in *Capital* reveals a great many different explanatory and descriptive efforts, and a number of distinct forms of analysis. Moreover, few of these regions of *Capital* are reducible to the others. None can be described as the definitive scientific content of the book, with the rest merely functioning as a body of implications or illustrations. Rather, Marx's account is irreducibly pluralistic in that it depends essentially on a variety of different forms of analysis and descriptive matter. And this plurality is in striking contrast to the deductive unity postulated by the naturalistic conception of theory.

Hypothesis and Description

Let us now turn to a second important dimension of the naturalistic interpretation: the view that scientific explanation generally proceeds through the postulation of unobservable entities, mechanisms, forces, structures, and so forth, which give rise to observable phenomena. According to this view, unified scientific theories are the vehicles through which such hypotheses are developed, and observable phenomena are explained by showing how they can be subsumed under the general features of the postulated mechanisms.

This view of scientific theories is widespread, though not universal, among adherents of naturalism within the philosophy of science generally.[26] It also has important proponents in contemporary treatments of Marx's science. Especially noteworthy are current efforts to apply the doctrines of scientific realism to his analysis of capitalism. In this vein Bhaskar, Keat and Urry, and others argue that Marx's system postulates unobservable mechanisms (generally abstract social structures) and that these mechanisms must be construed realistically. Thus Keat and Urry write, "[Marx] takes the more apparent and observable features of social life to be explicable in terms of these underlying structures. They can be comprehended by the discovery of the causal mechanisms central to each structure; these mechanisms are characterized in terms of the relations between a small number of theoretical entities. . . . He rejects the ideal of positivist science, the search for general laws and its connected model of explanation. He does however believe in the possibility of an objective science of social formations. He is both a naturalist and a realist."[27] Similar views can be found in the writings of Roy Bhaskar, Edward Nell, and many others.[28] According to these accounts, Marx explains phenomena in terms of underlying mechanisms, and these mechanisms are unobservable "theoretical entities" whose properties can be discovered only through extensive hypothetical and theoretical reasoning. This view of Marx's science thus shares most features of the naturalistic model of scientific knowledge and explanation. Thus the realistic interpretation of Marx's system is committed to the notion that hypothetical entities are central to his account of capitalism, and this commitment in turn suggests the "unified-theory" thesis. For how else are we to represent the properties of hypothetical entities but through organized systems of deductive laws and assumptions? The arguments that follow will undercut these antipositivist construals of Marx's science as well.

Against these views, I hold that Marx's explanations in *Capital* do not typically depend on hypothetical constructs or theoretical entities. Instead, his explanations depend on observable factors that can be independently examined using common tools of historical and empirical investigation. In explaining a given feature of capitalism Marx describes observable features of the capitalist economy and then shows how these features lead to the phenomenon under consideration through observable social processes or mechanisms. Thus in explaining the tendency of the rate of profit to fall, Marx singles out the individual capitalist's dominant motive – to maximize profits through technical innovation. He then shows that the aggregate result of large numbers of capitalists acting according to this incentive is a tendency for the rate of profit to fall. And the mechanism that connects explanans and explanandum is the process of competition among rational capitalists. Each premise of this explanation may be directly investigated through common tools of historical or economic investigation.

The broad distinction here is between theories that postulate the properties of hypothetical entities, on the one hand, and descriptive-analytic accounts that refer primarily to states of affairs that can be investigated directly using the research techniques of the discipline. The former approach explains observable events by deducing a description of the event from the laws postulated to govern the hypothetical entity or process. The postulated entity itself, however, can be investigated only indirectly through testing of the theoretical system as a whole. The latter account, on the other hand, explains an event by identifying an observable set of circumstances and showing how the event comes about as a result of those circumstances.

On this approach, Marx's main explanations are descriptive-analytical rather than hypothetical. Their chief premises can be defined and investigated without reference to assumptions unique to *Capital*. In particular, the concepts by which Marx characterizes the defining features of capitalism (class, property, technology, profit, wage, commodity, etc.) are substantially descriptive on this criterion – in marked contrast to concepts such as atom, gene, and gravity wave. These concepts are in common currency among social scientists and historians; it is possible to discuss them without committing oneself to the particulars of Marx's analysis of capitalism; and common tools of empirical and historical investigation permit any investigator to evaluate Marx's claims about these categories and their application to capitalist society.

This view gains support from a wide range of Marx's methodological writings. Thus in *The German Ideology* Marx writes, "The premises from which we begin are not arbitrary ones, not dogmas, but real premises from which abstraction can only be made in the imagination. . . . These premises can thus be verified in a purely empirical way" (*GI*, p. 42). And similar views are contained in his *Notes on Adolph Wagner* (*TM*, pp. 179–219), written in 1879–80. There Marx describes the logical character of his research. "In the first place I do not start out from 'concepts', hence I do not start out from 'the concept of value', and do not have 'to divide' these in any way. What I start out from is the simplest social form in which the labour-product is presented in contemporary society, and this is the 'commodity'. I analyse it, and right from the beginning, in the form in which it appears" (*TM*, p. 198). This passage reflects exactly the view outlined here: the concept of the commodity is not a theoretical construct but an observable category within capitalist society. The empirical characteristics of the commodity can be investigated independently of other assumptions in *Capital*. Marx's analytic work consists in working out the institutional framework within which commodities are produced and exchanged, and the institutional consequences to which these arrangements give rise.[29]

The descriptive nature of Marx's basic assumptions becomes even more evident when we consider particular examples of his explanations and analyses. (Chapter 5 will present a more extensive discussion of Marx's explanatory paradigm.) Such an examination shows that Marx's chief explanatory hypotheses generally have to do with descriptions of the constraining conditions within which economic agents act and the motivational factors that guide their behavior within the capitalist economy. For example, Marx describes the property system of capitalism as one in which a minority owns the means of production, while the majority owns only its labor power. "The second essential condition which allows the owner of money to find labour-power in the market as a commodity is this, that the possessor of labour-power, instead of being able to sell commodities in which his labour has been objectified, must rather be compelled to offer for sale as a commodity that very labour-power which exists only in his living body" (*Capital I*, p. 272). This is structural condition is of primary importance in Marx's account because it establishes the central conditions on motivation and strategies that constrain both parties (the capitalist and the worker). But this premise is not a highly theoretical assumption. Rather, this premise can be considered without regard to other assumptions at work in *Capital*, using tools of empirical research common to social science generally.

Similarly, to explain the tendency towards centralization of capital, Marx describes the conditions imposed on capitalist managers by the requirement of production for profit within conditions of competition. "The battle of competition is fought by the cheapening of commodities. The cheapness of commodities depends, all other circumstances remaining the same, on the productivity of labour, and this depends in turn on the scale of production. Therefore the larger capitals beat the smaller. . . . [This process] always ends in the ruin of many small capitalists, whose capitals partly pass into the hands of their conquerors, and partly vanish completely" (*Capital I*, p. 777). Given that capitalist firms are obliged to maximize profits by the conditions of competition and that capitalist managers are relatively rational and knowledgeable, Marx shows that capitalist firms tend to reinvest a portion of profits in expanded production. Again, what makes this account explanatory is not the unobservable quality of the motivational factors identified, but rather the discovery of the logic to which these factors give rise.

The most obvious objection to this treatment of Marx's system concerns the nature of the labor theory of value. Is value (socially necessary labor time) an observational concept? Is the rate of surplus value observable? In putting forward the labor theory of value and surplus value, Marx postulates that capitalist firms will be organized so as to maximize the rate of surplus value, and he holds that the rate of surplus value is not observable for the capitalist manager. Rather, the rate of profit is the observable category. Does this entail that Marx's system is theoretical in the strong sense after all?

The role of the labor theory of value in Marx's system will be discussed in greater detail in chapter 3. However, several relevant points can be considered in this context. First, Ernest Mandel maintains that value is in fact an observational concept in that it is a mathematical function of observable technical characteristics of the process of production. It would be possible in principle to collect the information about the composition of the process of production in all industries that is necessary to compute values for commodities, just as complex aggregate data are collected to determine the rate of unemployment and the rate of inflation.[30] Second, the claim that the rate of surplus value is unobservable to the capitalist manager is not quite correct. It is true that the rate of profit is the most pressingly observable category to him. But it would be possible for the capitalist to monitor the rate of surplus value as well, if appropriate statistical information were collected by some central agency. What is more relevant is that the capitalist manager has no interest in the rate of surplus value, since his own success or failure is determined by the behavior of the rate of profit. Third, and most fundamentally, it will be argued in chapter 3 that the labor theory of value is theoretically dispensable within Marx's system. On the account to be argued there, the labor theory of value is not a substantive empirical theory but an analytic apparatus in terms of which to aggregate and manipulate the basic economic characteristics of the capitalist economy. Other such instruments exist, however, that would serve Marx's explanatory purposes equally well and which do not refer to "embodied labor time." From this reasoning it follows that the labor theory of value is not an essential part of Marx's analysis of capitalism. Rather, what are essential are his assumptions about the institutional needs and incentives that define capitalist activity. And these assumptions are observational in the sense specified here.

These arguments suggest that the realism advanced by Bhaskar, Keat and Urry, and others is misguided when applied to *Capital*. Realism maintains that (1) the structures that are explanatorily fundamental to the phenomena of capitalism are real social entities, and (2) that these structures are theoretical constructs in that they can be known only through articulated theoretical systems. On the position taken here, however, the second premise of this interpretation is mistaken. Marx would certainly accept premise (1); he plainly believes that the structures and relations which he identifies are real social entities (e.g., the property relations, the production relations, the ideological structures). But these relations and structures are not theoretical constructs; they are not identified or investigated through a unified theoretical system but rather through a series of independent and fairly nontheoretical explanations and analyses. This qualification suggests that realism is flawed in precisely the feature it shares with naturalism, namely, the assumption that scientific knowledge ideally takes the form of unified theoretical systems.

It might be noted that this feature of Marx's system (its use of descriptive concepts rather than hypothetical ones) reflects to some degree a difference between natural science and social science more generally. Social phenomena are subject to a highly complex array of potentially relevant observable causal factors; the problem confronting the scientist is to determine which among these exerts the greatest influence. Thus the explanatory factors are hidden by their diversity rather than their unobservability.[31] The result is that the scientist must offer a hypothesis about which of the observable factors is explanatorily primary. But once selected, these factors are neither hypothetical nor theoretical.

Laws and Tendencies

A third important difference between Marx's system and the conception of theory described earlier concerns the status of the laws to which each refers and the ways in which the phenomena in question are thought to be lawgoverned. The paradigm of classical mechanics offers a simple and influential basis for answers to these questions: Events within a mechanical system are strictly governed by a small number of causal laws; those laws are universal and deterministic and they derive their necessity from fundamental properties of nature.[32] On this account, explanations in natural science depend upon laws of nature — exceptionless regularities that underlie and determine causal sequences. And the purpose of scientific research on this account is to formulate theories that uncover such laws and to show deductively how these laws give rise to the empirical phenomena within the domain of the theory.

As already noted, the naturalistic model is only partially accurate in application to actual examples in the natural sciences. But whatever its adequacy in natural science, this conception of a law of nature is largely unsuited to social science. The search for "iron laws" of society, or a small set of precise, universal laws that determine the bulk of social phenomena, is unlikely to bear fruit. As a consequence, the conception of social science as the quest for theories that formulate such laws is likewise unjustifiable. Social phenomena are indeed law-governed, but they are so in a way very different from natural phenomena.

In particular, *Capital* does not formulate anything seriously analogous to strict laws of nature. This point will perhaps seem paradoxical, since Marx describes his aim as that of discovering the "laws of motion" of the capitalist economy and was frequently drawn to the example of celestial mechanics. In a celebrated passage in the Introduction to *Capital* he writes, "It is the ultimate aim of this work to reveal the economic law of motion of modern society" (*Capital I*, p. 92). And later in *Capital* he explicitly casts his investigation in analogy with celestial mechanics: "This much is clear: a scientific analysis of competition is not possible, before we have a conception of the inner nature of capital, just as the apparent motions of the heavenly bodies are not intelligible to any but him, who is acquainted with their real motions, motions which are not directly perceptible by the senses" (*Capital I*, p. 433). However, the idea of a law of motion of society must be understood in a very loose sense. Marx certainly regards the CMP as a law-governed system; the question is, what sort of laws do the governing? Both common sense and the details of Marx's own analysis support the conclusion that these laws are fundamentally different from laws of nature. The CMP is not a "natural" system, and its regularities arise not from natural necessity but from a "logic of institutions" through which the CMP constrains individual actions and imposes a pattern of organization and development on social institutions more generally.

A clear indication that Marx does not regard the laws governing the capitalist mode of production as laws of nature can be found in his insistence on the historical particularity of these laws and his criticisms of Mill, Ricardo, and others on this point. "Whenever we speak of production, then, what is meant is always production at a definite stage of social development" (*Grundrisse*, p. 85). Thus the laws governing the capitalist mode of production are historically specific; that is, they are different from those laws that govern feudal or slave production. And a fundamental error of classical political economy is its effort to represent the laws of capitalist production as eternal laws of nature: "The aim [of the political economist] . . . is to present production . . . as encased in eternal natural laws independent of history, at which opportunity bourgeois relations are then quietly smuggled in as the inviolable natural laws on which society in the abstract is founded" (*Grundrisse*, p. 87). Similar views occur in *The Poverty of Philosophy*: "When the economists say that present-day relations – the relations of bourgeois production – are natural, they imply that these are the relations in which wealth is created and productive forces developed in conformity with the laws of nature. These relations therefore are themselves natural laws independent of the influence of time. They are eternal laws which must always govern society. Thus there has been history, but there is no longer any" (*PP*, p. 121).

These criticisms plainly imply that the laws Marx identifies are not laws of nature but rather historically specific social laws.

Marx uses the idea of a "law of tendency" to represent the fact that the regularities he deduces fall far short of the universality and precision of laws of nature. For example, in discussing his assumption that the rate of surplus value is constant across all industries, he writes, "We assume a general rate of surplus-value . . . as a tendency, like all economic laws, and as a theoretical simplification; but in any case this is in practice an actual presupposition of the capitalist mode of production, even if inhibited to a greater or lesser extent by practical frictions that produce more or less significant local differences" (*Capital III*, p. 275). Thus Marx characterizes the law of the constancy of the rate of surplus value—and implicitly, all other economic laws—as a law that formulates a tendency within the capitalist mode of production that may be offset by a multitude of different factors.

Marx's argument for the falling tendency in the rate of profit provides a clear illustration of his conception of laws of tendency. He shows that the drive toward accumulation induces each capitalist to invest in new capital-intensive technologies, resulting in a tendency toward rising capital-labor ratios. (The argument for this conclusion will be presented in greater detail in chapter 5.) Given, however, that profits are determined by the amount of surplus value created, and that surplus value is produced in proportion to the value of the wage, this tendency leads to a fall over time in the rate of profit. This argument therefore provides an explanation of the known tendency toward a falling rate of profit.[33] However, Marx also recognizes that other factors at work in the economy tend to offset this falling tendency. "Counteracting influences must be at work, checking and cancelling the effect of the general law and giving it simply the character of a tendency, which is why we have described the fall in the general rate profit as a tendential fall" (*Capital III*, p. 339). A few pages later he considers the fact that a rise in the rate of surplus value will offset the drop in the rate of profit. But he holds that "[this factor] does not annul the general law. But it has the effect that this law operates more as a tendency, i.e. as a law whose absolute realization is held up, delayed and weakened by counteracting factors" (*Capital III*, pp. 341–42). In his discussion of the "counteracting influences" that offset the falling tendency, Marx recognizes full well that the factors influencing the rate of profit are various, and that they work through the structuring of individual choices. Perceiving the fall in the rate of profit, the capitalist takes action to offset it (e.g., by speeding up the process of production, extending foreign trade, or reducing the wage), and these incentives for the individual eventuate in an upward tendency in the movement of the rate of profit.[34] Thus the regularity—that the rate tends to fall—is not strictly universal or necessary, but rather can be offset by contrary tendencies.

The law of the falling rate of profit is one of the clearest and most rigorously derived of the laws of motion which Marx identifies. Consequently his unambiguous recognition that this law is a "law of tendency" rather than an iron law of nature establishes that the laws of motion of the CMP to which he refers are not akin to laws of nature. Rather, the laws of motion are laws of tendency that derive from the incentives and constraints established by the capitalist economic structure for the agents (in this case the capitalist).

What are the logical characteristics of laws of tendency, and how do they differ from laws of nature? A law of tendency is a statement to the effect that, holding other background conditions fixed, condition C tends to give rise to pattern P. Examples of such laws include these: "Capitalist ownership tends to concentrate over time"; "The rate of profit tends to fall over time"; "The capitalist economy tends toward crises of underconsumption over time." Significantly, virtually all the law statements Marx uses or derives take the form of laws of tendency.

The notion of a law of tendency imposes at least two restrictions on the regularities it describes. First, it reflects the fact that social regularities generally have substantial ceteris paribus conditions, that is, assumptions postulating that relevant background conditions remain unchanged. For example, consider this social regularity: If the price of a good rises, consumption tends to fall. This statement is true only if one assumes that background conditions are normal and unchanging; so, for example, the regularity will fail to occur if the vendor expends more money on advertising to prop up sales or if demand is inelastic.[35] The ineliminability of ceteris paribus conditions is characteristic of causal claims generally (see, for example, J. L. Mackie's treatment),[36] but they play a more prominent role in social science than in natural science because social phenomena are subject to a wider range of causal influences. Even simple economic facts display this open character of social causation. The price of wheat grown in Minnesota may be influenced by climatic conditions in the Ukraine, disruption of oil supplies in the Middle East, sudden increases in public anxiety about the health effects of pesticides, shocks in the financial and commodities futures markets, movement in the prime interest rate, or widespread destruction of livestock due to disease. It is therefore impossible to determine a narrow range of factors that might be said to be causally decisive. (Weber addresses this point *The Methodology of the Social Sciences*.)[37] Consequently, whereas it is often feasible to isolate a physical system and artificially hold the ceteris paribus conditions fixed, this is seldom possible in social science.

Second, laws of tendency are limited because they derived from the actions of independent individuals, so that even if the ceteris paribus conditions are completely satisfied, the outcome is only partially predictable. Thus it is evident that the regularity with which price rises tend to result in reduced consumption still depends on an assumption of perfect consumer rationality, even if all ceteris paribus conditions are satisfied. Given imperfections of consumer rationality, however, the relation between price and consumption will not generally be the simple one predicted by the regularity statement. Even in the best case, therefore, laws of tendency are less precise and less predictively accurate than laws of nature.

These limitations on the precision and scope of social regularity statements reflect a basic difference in the underlying nature of the regularity: Laws of nature are thought to arise from the determinate properties of natural processes and mechanisms, whereas social laws derive (directly or indirectly) from facts about individuals making decisions within conditions of constrained choice.[38] Consider several examples of social regularities: "Capitalists seek to maximize profits"; "Consumption of a commodity tends to fall in response to rising prices"; and "Voter turnout increases in proportion to the intensity of public concern for campaign issues." In each case the regularity depends ultimately on the choices individuals make given a set of purposes and constraints. Thus social regularities are not based on natural necessity at all, and the metaphor of "iron laws" is wholly inappropriate. It seems to promise the sort of precision afforded for astronomy by celestial mechanics. But this precision is precluded by the very nature of the regularities inherent in social phenomena.

Marx's view of the mechanism by which the rate of surplus value tends to increase illustrates this dependency of laws of tendency on aggregates of individuals making independent choices. "When an individual capitalist cheapens shirts, for instance, by increasing the productivity of labour, he by no means necessarily aims to reduce the value of labour-power and shorten necessary labour-time in proportion to this. But he contributes towards increasing the general rate of surplus-value only in so far as he ultimately contributes to this result" (*Capital I*, p. 433). The individual manager reorganizes the process of production so as to reduce the unit cost of production for his own reasons – to permit him to realize a greater quantity of profit per unit. But the aggregate result of this incentive is a tendency toward a rising rate of surplus value within the economy as a whole through the cheapening of wage goods and a resulting drop in the value of labor power.

It must be acknowledged that there is some tension in *Capital* between the "iron-laws" interpretation of social regularities and the "laws-of-tendency" interpretation. Thus consider the following passage in which Marx refers to the laws of motion of capitalist society: "It is a question of these laws themselves, of these tendencies winning their way through and working themselves out with iron necessity" (*Capital I*, p. 91). This passage represents both interpretations in that it refers both to tendencies and to iron necessity. On balance, however, the "laws-of-tendency" model preponderates in the detailed explanatory work of

Capital. Marx's reasoning for the laws of motion he derives always leaves room for countervailing tendencies that may impede or reverse the working of the law. Indeed, this passage itself implicitly favors the law-of-tendency interpretation, since the metaphor of tendencies "winning their way" suggests the soft conception of social causation. Thus Marx's explanatory practice favors the laws-of-tendency interpretation, whereas his explicit comments about the significance of his findings occasionally lean toward the stronger iron-laws interpretation. Throughout this book, however, I have given primacy to Marx's practice as a scientist over his own interpretation of his work; and this issue should be treated similarly.

Logic of Institutions

These observations make it plain that the naturalistic interpretation of Marx's system is unsatisfactory: His construction is not a unified deductive theory in which all aspects relate back to the central hypotheses. Instead, I will hold in this section that Marx's system can be broadly divided into two parts. First, he describes the structural and functional characteristics that distinguish capitalism from other modes of production and that create the institutional matrix within which participants act. This account has historical and sociological content, but its chief purpose is to identify and describe the economic structure of capitalism—the social relations of production. Second, Marx constructs a large number of arguments purporting to determine the institutional implications of these features of the economic structure—the "laws of motion" of the capitalist economy. This part of Marx's analysis is intended to establish the "institutional logic" imposed by the underlying institutions of capitalism on the development and organization of the economy.

Descriptive Content of *Capital*

Much of Marx's work in *Capital* is an effort to identify the structural and functional properties of the capitalist mode of production that distinguish it from other modes of production. This work is largely descriptive; it depends chiefly on accurate observation of the phenomena of capitalism rather than complex theoretical analysis of the capitalist economy.

The aim of description in science is to provide an empirically accurate representation of a range of phenomena, and that task is of the greatest importance in fields of science in which the phenomena themselves are highly complex and unwieldy. Any range of phenomena is a richly textured whole, with indefinitely many properties. It is therefore possible to describe social phenomena from a correspondingly diverse range of points of view. The first task for scientific description, therefore, is the selection of a set of factors as being of particular explanatory interest. These considerations show that scientific description generally proceeds on the basis of a background hypothesis about the causal workings of the phenomena in question. However, the adequacy of a scientific description can be assessed independently of the explanatory hypothesis that motivates it using straightforward criteria of empirical evaluation.

Historical materialism provided Marx with a basis by postulating that factors having to do with the economic structure are of primary explanatory import. Consequently his descriptive work in *Capital* proceeds within a larger explanatory hypothesis; it characterizes capitalism in terms of the institutions that define its economic structure. However, having used the theory of historical materialism to single out the economic structure for close scrutiny, the accuracy of Marx's account does not depend on the correctness of the full theory of historical materialism. If we ultimately reject the explanatory theses of historical materialism, we can conclude that a detailed depiction of the economic structure is of little interest but that the accuracy of that account can be assessed independently.

Marx's chief descriptive contributions include these:

1. A description of the system of property.
   a. Private ownership of means of production.
   b. Wage labor (labor power is a commodity).
2. A description of the purpose of production.
   a. Production of goods for exchange (commodity production).
   b. Production for profits.
   c. Profits used for accumulation rather than luxury consumption.
3. A description of the working of a competitive commodity-producing system.
4. A sociological account of how these conditions are reproduced.
5. A description of various aspects of the process of production: the labor process, the technical and social division of labor, and the effects of capitalist production on the working class.

The first set of characteristics included here defines the distinctive structural features of the economic structure of capitalism: the property relations within which capitalist economic life takes place. Marx analyzes modes of production in terms of the form of surplus extraction each represents. Every society has some set of social relations of power and authority through which a surplus is extracted from the immediate producers, and these social relations define the property system. Different modes of production represent distinct techniques of expropriation. Thus slavery depends on forcible surplus extraction; that is, the slave is compelled by threat of violence to labor for the master beyond the amount of time needed to reproduce the slave's means of life. Feudalism depends on a network of traditional, religious, and military relations to compel the serf to labor for the lord. And capitalism depends on the separation of the worker from the means of production. In order to live, the worker is compelled to sell labor power as a commodity; and in general the value of this labor power is less than the length of the workday. The excess labor time represents surplus labor.

Thus the property relations that define a mode of production structurally are the relations of effective power through which the forces of production are utilized, and on the basis of which the surplus product is distributed and used. These conditions distinguish capitalism from other systems of production, for example, feudalism, classical slavery, and freeholding agriculture. They also define the class nature of the mode of production, since class membership is defined in terms of position within the property system.[39]

The capitalist economic structure is defined by the relations of private property. Each individual buys and sells commodities and has extensive freedom in the use and disposition of these commodities. "[The capitalist system] arises only when the owner of the means of production and subsistence finds the free worker available, on the market, as the seller of his own labour-power. And this one historical pre-condition comprises a world's history. *Capital*, therefore, announces from the outset a new epoch in the process of social production" (*Capital I*, p. 274). A minority class (the bourgeoisie) owns and controls the means of production (machinery, land, mines, transportation and communication facilities, buildings), while the rest of society (the proletariat) owns only its labor power. "This worker must be free in the double sense that as a free individual he can dispose of his labour-power as his own commodity, and that, on the other hand, he has no other commodity for sale, i.e. he is rid of them, he is free of all the objects needed for the realization of his labour-power" (*Capital I*, pp. 272–73). The capitalist system depends, therefore, on separating the direct producer from the means of production. Because the members of the proletariat are denied free access to the means of production (e.g., in the form of common lands or socially owned factories and tools), they are forced to sell the commodity they possess in order to buy the means of life; they must become wage laborers and sell their labor power. The economic structure of capitalism thus embodies a class division between those who own productive capital and those who are forced to sell their labor power.

Much of *Capital* describes these structural features of capitalism. Marx discusses the distinctive features of bourgeois property in chapter 6 ("The Sale and Purchase of Labour-Power," pp. 270–80); he demonstrates the necessary relationship between class and profit in chapters 7 and 8 (esp. pp. 293–306); and he describes the historical process through which these structural conditions came into being in his account of primitive accumulation (pp. 873–930). He describes the implications of the ownership of capital in chapter 24 (esp. pp.

725–34), and the impact of these property relations on the worker in chapters 10, 14, and 15 ("The Working Day," "The Division of Labour and Manufacture," and "Machinery and Large-Scale Industry").

The second set of characteristics defines the functional properties of the capitalist mode of production. These conditions describe the *purpose* of capitalist production. Goods are produced in every economy, but capitalism is a system of commodity production, that is, production of goods for exchange. This feature is historically specific, since the feudal economy, classical slavery, and peasant agriculture are not primarily organized around exchange. Peasant agriculture, for example, is organized chiefly around consumption.[40]

Production for profit further distinguishes capitalism from other commodity-producing economies. In a simple commodity-producing system, independent producers create commodities for exchange as use values. The tailor produces shirts full time in order to exchange them on the market for shoes, food, tools, and other necessities of life. Since these latter commodities differ qualitatively from the shirts the tailor produces, the exchanges are rational without presupposing a net increase in value. In capitalism, however, production is organized around an increase in the total amount of wealth available to the capitalist. The capitalist makes an investment of his wealth in order to earn a profit on his investment. Marx represents this schematically in the form M-C-M′: The capitalist advances money by purchasing commodities (constant and variable capital), produces new commodities, and sells these new commodities for a new (and larger) quantity of money (*Capital I*, pp. 247–57). This series of transactions is plainly irrational if the quantity of money eventually received is equal to that originally advanced; so the receipt of profit on investment is thus intrinsic to capitalist economic organization.

Capitalism is further distinguished from other profit-based systems in that it is organized for accumulation rather than luxury consumption. In contrast to an economy in which the propertied class spends the surplus it extracts in luxury consumption, capitalism requires that the capitalist reinvest a portion of profits in expanded production.

The final presupposition of capitalist production is a system of competitive market relations through which incomes are distributed and on the basis of which investment decisions are made. Each agent is assumed to act so as to rationally maximize income: Workers are concerned to increase their salaries and capitalists their profits.

The structural and functional features of capitalist production are interdependent. Production for profits is possible only within a system of property in which labor power is bought and sold, namely, a system of wage labor. Such a system depends on a division between those who own the means of production and those who own only their labor power. Production for profit and accumulation therefore depends on a particular economic structure: a system of private property in means of production and labor power.

These functional requirements of capitalist production are emphasized throughout *Capital*. The circuit of capital (M-C-M´) that characterizes capitalist production is extensively analyzed in chapter 4 ("The General Formula of Capital"). The logic of accumulation is discussed in chapters 24 and 25; the coercive requirement of maximizing profits is described in chapters 10 (esp. pp. 340–44) and 12; and the logic of competition is recognized throughout (e.g., in Marx's discussion of the requirement of equal rates of profit in different industries; *Capital III*, chapters 9 and 10).

Institutional Logic

Two directions for scientific inquiry arise once Marx has identified these distinguishing features of capitalist production. First, we might pose a set of historical and sociological questions concerning the genesis and reproduction of these factors. How did bourgeois property come to replace feudal property? How was a landed peasantry effectively dispossessed of its rights of access to the land? How is the separation of propertied and nonpropertied persons maintained in ongoing capitalism? As Marx puts it, "One thing, however, is clear: nature does not produce on the one hand owners of money or commodities, and on the other hand men possessing nothing but their own labour-power. This relation has no basis in natural history. . . . It is clearly the result of a past historical development, the product of many economic revolutions, of the extinction of a whole series of older formations of social production" (*Capital I*, p. 273). Thus it is a theoretically important question to investigate the social processes by which these institutional presuppositions of the capitalist system came into being. Marx considers this sort of question at various points in *Capital*, and particularly so in his discussion of primitive accumulation.[41]

Second, however, we may ask what the institutional implications are of these conditions for the organization, development, and reproduction of the capitalist economy. For an economy is a system of institutions in which various parts must fit together in order for the economy to work adequately, and it is defined by a set of social relations that both motivate and constrain the activities of individuals pursuing goals within the economic system. "The capitalist process of production, therefore, seen as a total, connected process, i.e. a process of reproduction, produces not only commodities, not only surplus-value, but it also produces and reproduces the capital-relation itself; on the one hand the capitalist, on the other the wage-labourer" (*Capital I*, p. 724). A central concern in much of *Capital* is Marx's effort to discover systemic consequences of specific features of capitalist social relations of production. He believes that different modes of production have distinctive historical "signatures" and that their distinctive laws of motion follow from the fundamental properties of their economic structures. Applying the analytic tools of political economy to his study of the defining features of the capitalist economic structure, Marx attempts to work out the institutional logic of these capitalist institutions. What distinctive features of organization and development are imposed on the capitalist economy by its defining structural and functional characteristics? What are the "laws of motion" of the mode of production defined by these conditions?

We may call this an institutional-logic analysis of social regularities, and it is significantly different from the construction of theoretical explanations in natural science. Such an analysis is concerned with determining the results for social organization and development of an entrenched set of incentives and constraints on individual action. Marx gives a concrete example of such a logic when he refers to the revolutionary implications of innovations in one area of the economy for other areas. "The transformation of the mode of production in one sphere of industry necessitates a similar transformation in other spheres. This happens at first in branches of industry which are connected together by being separate phases of a process, and yet isolated by the social division of labour, in such a way that each of them produces an independent commodity" (*Capital I*, p. 505). Putting the idea fairly simply, social changes and patterns result from the actions of many different human agents acting out of a variety of motives. These individuals are subject to very specific incentives and conditions that limit their actions and propel them in particular directions. Large-scale social tendencies can be explained in terms of the conditions within which individuals plan and act. Such an explanation proceeds from a description of the conditions under which individual participants make their choices, and arrives at a conclusion about the overall consequences which follow from these conditions for the system as a whole.

A simple example of this sort of pattern may be seen in any university green: the shortest path between two popular points is generally bare of grass, as each individual seeks to save a small amount of time by taking a shortcut. No individual singly flattens all the grass, but the end result is the creation of a bare path. Similarly, a particular system of property creates specific conditions within which individual activity takes place. It creates distinctive incentives and prohibitions that structure individual conduct, and it creates institutional "needs" that stimulate the emergence of other institutions in their turn to satisfy these needs.[42]

On this view, explanation of a social fact P consists of an idealized description of a set of social institutions, and a demonstration of how P emerges as a result of individuals pursuing their goals within those institutions.[43] The criteria of adequacy of such an analysis are fairly clear: It should be detailed and specific; the arguments offered purporting to establish connections between factors should be rigorous; and attention must be paid to the fullness of available empirical evidence. Such an analysis does not usually rest on hypotheses about unobservable mechanisms, and it does not take the form of a unified deductive system.[44] Finally (as will be considered more fully later in this chapter), such an account is justified by assessing the descriptive accuracy of its premises and the rigor of its reasoning in establishing the institutional logic it describes. (Institutional-logic explanations will be considered in greater detail in chapter 5.)

This analysis of Marx's system is closely parallel to recent efforts to provide microfoundations for Marxist theory. Robert Brenner, Jon Elster, John Roemer, and others have put forward theories that attempt to connect the general laws formulated in Marx's system with the constraints on individual action embodied in the capitalist system. Thus Brenner writes, "I began from the idea that social-property systems, once established, tend to set strict limits and impose certain overall patterns upon the course of economic evolution. They do so because they tend to restrict the economic actors to certain limited options, indeed quite specific strategies, in order best to reproduce themselves – that is, to maintain themselves in their established socio-economic positions." [45] Jon Elster takes this form of analysis in a slightly different direction by attempting to apply technical results from the theory of games to the reasoning from the individual's circumstances to the collective outcome. "Game theory is invaluable to any analysis of the historical process that centers on exploitation, struggle, alliances, and revolutions." [46] Finally, John Roemer's important work in the theory of exploitation reflects a similar orientation; his project is to uncover the institutional characteristics necessary for a social system to be exploitative.[47] Each of these authors attempts to provide a microlevel foundation for Marx's general conclusions by describing in detail the institutional arrangements of a capitalist society, and by showing how individuals acting rationally within the context of these institutions will give rise to distinctive collective patterns.[48]

According to this interpretation of *Capital*, Marx provides a selective but detailed description of the economic structure of capitalism and offers a rigorous analysis of the institutional implications of these features for the economy as a whole. To the first end, he delineates the structural and functional characteristics that distinguish capitalism from other modes of production. This portion of Marx's work is largely descriptive. And toward the second end, he provides an abstract economic analysis of capitalism that attempts to work out the laws of motion defined by these characteristics. Here Marx's intention is to establish the institutional logic – the "laws of motion" – defined by the underlying institutions of capitalism.[49]

This form of analysis can be seen clearly in Marx's explanation of the industrial reserve army. Each capitalist has an incentive to reduce his own labor costs; this incentive becomes stronger as the wage rises, and overall the result is that rising wages result in widespread innovations reducing demand for labor. These factors lead in the aggregate to the creation of a pool of unemployed and underemployed workers. In other words, a global pattern — persistent unemployment — follows from the structure of incentives governing each capitalist.

This account squares with Marx's portrayal of the capitalist as the creature of his social position. "I do not by any means depict the capitalist and the landowner in rosy colours. But individuals are dealt with here only in so far as they are the personifications of economic categories. . . . My standpoint . . . can less than any other make the individual responsible for relations whose creature he remains" (*Capital I*, p. 92). This view of the relation between economic laws and individuals is given more detail later in *Capital*: "Capital developed within the production process until it acquired command over labour, i.e. over self-activating labour-power, in other words the worker himself. The capitalist, who is capital personified, now takes care that the worker does his work regularly and with the proper degree of intensity" (*Capital I*, p. 424). The capitalist stands within a set of incentives and prohibitions that severely restrict his freedom of action. The incentives and limits that determine the actions of the bourgeois manufacturer are substantially different from those of the feudal lord. And the capitalist who flouts the most central of these conditions will not long remain a capitalist. This fabric of motivational constraints gives rise to distinctive collective patterns for the capitalist system as a whole. At the same time, however, economic laws and categories have no necessity beyond the incentives and prohibitions imposed on individual actors and the collective patterns that emerge from these incentives.

The "logic-of-institutions" model allows us to make sense of the features of Marx's system that are most inconsistent with the naturalistic interpretation. First, the "institutional-logic" model is perfectly consistent with the irreducible plurality of explanatory and descriptive components of Marx's account. Some of the components identified earlier can be classified as basic institutional features that give rise to such a logic; others represent tendencies that result from these features; and others take the form of historical comments on the genesis of such institutional features. Whereas the naturalistic model favors a tightly unified system, the institutional-logic model tolerates a looser form of analysis and description.

Second, the idea of a logic of institutions provides an underlying metaphysics for the idea of a law of tendency. It establishes the motive power that stands in place of laws of nature and natural necessity in Marx's explanations. If we ask why the rate of profit falls, the answer to which we are ultimately led is this: Capitalists rationally strive to maximize their rate of profit, and the strategies available to them lead in the long run to a downward tendency in the rate of profit. Thus the circumstances of choice that constrain the capitalist manager impose a logic on the economy as a whole.

Finally, the fact that an analysis of the logic of capitalist institutions depends largely on observable features of capitalist society explains the absence of hypothetical entities in Marx's system. Explanations within such an analysis depend on successful identification of the causally relevant institutional conditions, not on theoretical hypotheses about underlying mechanisms.

Contemporary Marxist Social Science

One useful way of assessing the adequacy of this interpretation of the logic of Marx's social science is to see how it accords with the practice of contemporary Marxist social scientists. Even a superficial review of this literature yields a good deal of support for the account being advanced here. For on the whole, Marxist social scientists in the past two decades have *not* made use of Marx's theory as a closed deductive system encapsulating the "essential properties" of all capitalist systems. They have not approached *Capital* as a body of social laws that can be used to predict and explain in detail the contemporary workings of capitalist society. And they have not regarded their work as the simple application of a finished theoretical system to the data that have become available in the past 120 years since the publication of *Capital*. In a word, Marxist social scientists have not treated Marx's analysis of the capitalist mode of production as a naturalistic body of theory, concepts, and laws to be more or less mechanically applied to contemporary capitalist society, and they have not attempted to provide such theories themselves.

Let us consider several important examples. In *State in Capitalist Society*[50] Ralph Miliband elaborates the ways in which economic class interest can be transformed into political power within modern capitalist society, and the ways in which the form and policies of the capitalist state are molded to suit the needs of the property system. His primary debt to Marx's system has to do with the theory of class. He takes as given the basic outlines of Marx's analysis of the property arrangements characteristic of capitalism (pp. 23–48) and the system of classes to which these property arrangements give rise. With these theoretical ideas in the background, Miliband examines the institutional constraints through which a property system with these characteristics produces particular state forms and policies. Much of the value of his treatment, moreover, is its empirical detail. Miliband's work thus reflects several aspects of the earlier account: He is concerned with identifying an institutional logic (the pattern of political development brought about by the property arrangements of capitalism); he submits his analysis of this logic to appropriate empirical constraints; and he regards

Marxist "theory" as a loosely organized collection of insights into various aspects of capitalism, rather than a finished specification of the necessary workings of every capitalist system.

Samuel Bowles and Herbert Gintis represent a second important example of contemporary Marxist social science. Their *Schooling in Capitalist America*[51] is a study of the role educational institutions play in reproducing the social structures of capitalist society. Their work too is deeply indebted to Marx's analysis. They take over much of Marx's theory of class and his account of the system of property that characterizes capitalism (pp. 57–68); and they make special use of Marx's view of the centrality of work relations within the system of production. But Bowles and Gintis also recognize that it is not possible to deduce the role of educational institutions from Marx's analysis of capitalism. Rather, it is necessary to investigate the specific empirical circumstances of those institutions and discover their relations to other aspects of capitalist society. Classical Marxist theory offers suggestive hypotheses about the role of these institutions (e.g., in the theory of ideology). Bowles and Gintis substantially extend this extremely schematic account, both empirically and analytically. And they are primarily concerned to analyze some of the institutional means by which the needs of the system of property are imposed on the system of education.

Finally, consider Harry Braverman's *Labor and Monopoly Capital*,[52] the most closely connected to *Capital* of any work surveyed here. In effect Braverman undertakes to update the treatment of technology and work relations in *Capital*. But Braverman does not take the position that the theoretical accomplishments of *Capital* suffice to allow us to deduce the nature of the social relations of twentieth-century capitalism. Quite the contrary, his detailed empirical discussion of the labor process of modern capitalist production adds a great deal to our knowledge of modern capitalism. And this knowledge could not in any fashion be deduced from the theoretical assumptions of *Capital* alone. Thus with Braverman too we find that the research proceeds within the general theoretical assumptions worked out in *Capital* but that he recognizes full well the necessity to extend, modify, or adapt these theoretical assumptions within the context of available empirical data.

Thus in each of these cases we find that Marxist social science does *not* treat Marx's "theory" of capitalism as a canonical work that establishes the necessary properties of all capitalist systems. Instead, Marxist social scientists have recognized that contemporary capitalist society is in many important ways different from the mode of production Marx studied (and, naturally, similar in many ways as well). They have identified different ranges of phenomena for investigation than those Marx considered. And they have put forward theoretical analyses of various aspects of capitalist society that go significantly beyond Marx's own formulations.

What these social scientists have in common with Marx is a fund of central concepts—the concepts of class, property, technology, mode of production, and the like—and a loose set of shared hypotheses about what sorts of factors are explanatorily fundamental within capitalist society. These scientists are working within a program of research largely defined by Marx's investigations of capitalist society.[53] But their work cannot be construed as the mere articulation and development of a tightly organized theory as expounded in *Capital*. Rather, the social scientist takes over some of Marx's theoretical assumptions and works these into an original examination of various aspects of contemporary society. What emerges is a detailed analysis of specific empirical phenomena, given theoretical organization by some of the concepts and explanatory assumptions of classical Marxist theory. But these theoretical assumptions alone do not suffice to determine the content of the analysis. This feature of current Marxist social science corresponds exactly to the account provided earlier of the logical organization of Marx's own study of different aspects of the capitalist economic structure.

Significantly, the assumptions shared by all the authors surveyed are those previously identified here as being at the core of Marx's study of the basic institutions of the capitalist economy: the property relations of capitalism, the class system created through those relations, and the set of incentives and interests imposed by the class system on the men and women who live within those social relations. On the basis of these assumptions, these authors provide original analyses of aspects of the modern capitalist economy that are not considered in detail in *Capital*. And they typically proceed on the basis of an account of a "logic of institutions" rather than any naturalistic conception of direct causation.

Thus a brief survey of some contemporary Marxist empirical social science lends further support to the basic view advanced here of the logic of Marx's own examination of capitalism—not as a finished, unified deductive theory which permits the derivation of all relevant particulars of capitalist society, but rather as a set of related explanatory hypotheses that can be used flexibly in constructing an analysis of particular phenomena of capitalist society.

2

Historical Materialism and *Capital*

Marx's economic research was located within a larger program of research: historical materialism. Historical materialism provided both the general problem of investigations for Marx's economics (to work out the essential features of the economic structure of capitalism) and the general concepts in terms of which Marx defines his project (economic structure, production relations, productive forces, mode of production, class, property, and so forth). Moreover, historical materialism led Marx to conceive of social explanation in terms that were substantially different from those of classical political economy by leading him to emphasize the historical specificity of social laws of organization and development. Further, Marx plainly takes much of the content of historical materialism as fixed in *Capital* (e.g., that ideology conforms to class interest). Thus historical materialism is essential to understanding the research goals and presuppositions Marx brings to his economics. This chapter presents some of the chief ideas of historical materialism and some of the conceptual problems that confront it.

The importance of historical materialism in guiding Marx's research notwithstanding, I will argue that his analysis of capitalism is to an important degree independent from the larger claims of his theory of historical materialism. *Capital* is not logically committed to many of the hypotheses that define historical materialism: the theories of ideology, politics, culture, and the general theory of the mode of production. It is, rather, a specialized study of a particular feature of capitalist society, namely, its economic structure. Moreover, historical materialism provided little help in the actual conduct of research in Marx's economics, and it did not supply the specialized theoretical tools he came to employ in analyzing the economic structure. The resources of historical materialism are not by themselves sufficient to produce a scientific examination of capitalism. The chapter concludes with an argument to the effect that historical materialism is not a scientific theory of history at all, but rather a program of research.

The Thesis and Chief Concepts

The Thesis

Marx formulates the theses of historical materialism in a number of different ways throughout *The German Ideology*, *The Communist Manifesto*, *The Holy Family*, and other works of the 1840s. Recent commentators have tried to provide an interpretation of the theory that brings the alternative statements into one coherent view.[1] Stated very schematically, historical materialism maintains that facts about the social and technical properties of the production process in a society—the forces and relations of production—"determine" the properties of noneconomic institutions, that is, the state, ideology, religion, and the like. (A central issue, however, concerns the nature of the concept of determination which is in use here.) The forces of production can be described as the forms of technology existent within a given society on the basis of which the society produces the goods needed to satisfy the needs of its population; the relations of production are the social relations of power and authority through which the forces of production are utilized and enjoyed. According to historical materialism, the system of production underlies and "determines" the noneconomic institutions and arrangements of society: In order to explain the details of the form of organization of the state or the dominant system of ideology or the prevalence of a certain type of religious institution, it is necessary to discover the particular features of the system of production that give rise to these superstructural arrangements. The thesis of historical materialism functions both as a quasi-empirical assumption, in that it supposes a certain contingent relationship between levels of social structure, and as methodological directive. It leads Marx to orient his research in a particular direction: toward a detailed analysis of the workings of the system of production.

The preface to *A Contribution to the Critique of Political Economy* provides a statement of the view that is often taken as its basis:

> In the social production of their existence, men inevitably enter into definite relations, which are independent of their will, namely relations of production appropriate to a given stage in the development of their material forces of production. The totality of these relations of production constitutes the economic structure of society, the real foundation, on which arises a legal and political superstructure and to which correspond definite forms of social consciousness. . . . At a certain stage of development, the material productive forces of society come into conflict with the existing relations of production or—this merely expresses the same thing in legal terms—with the property relations within the framework of which they have operated hitherto. From forms of development of the productive forces these relations turn into their fetters. Then begins an era of social

revolution. The changes in the economic foundation lead sooner or later to the transformation of the whole immense superstructure. (*EW*, pp. 425–26)

This passage states the thesis of historical materialism very clearly. The foundation of society is its economic structure – the "totality of these relations of production." The relations of production correspond to the stage of development of the material forces of production; they serve to develop the forces of production. Corresponding to a particular economic structure there develops a superstructure of law and consciousness, which supports the existing economic structure. As the forces of production develop, they come into conflict with the relations of production (or property relations). The relations of production become fetters and are more or less rapidly transformed into a new set of property relations more compatible with the new forces of production. Finally, as the new economic structure emerges, the superstructure changes to reflect more adequately the new economic relation. Schematically, the forces of production expand and burst the relations of production within which they developed, and the superstructure in turn changes in light of the modifications in the economic structure.

These theses constitute the heart of the theory of historical materialism.[2] It should be noted, however, that this formulation obscures or ignores a number of problems that have come to the fore in recent work in Marxist theory, for example, the nature of historical determination, the relative primacy of the forces and relations of production, the extent to which institutions of the superstructure may causally influence the economic structure, and the role of class conflict and politics in the historical process. Many of these problems will be considered later in this chapter. In spite of these shortcomings, however, the version set forth in the preface to *A Contribution* has the advantage of theoretical economy and simplicity, and serves well as a preliminary statement of the primary ideas of historical materialism.

Let us consider more carefully the main concepts of historical materialism: the forces and relations of production, the mode of production, and the economic structure.

The Chief Concepts

In the preface to *A Contribution to the Critique of Political Economy*, Marx describes the historical process in terms of the conflict between the forces and relations of production. These concepts jointly encompass both the technical and the social aspects of the system of production. They correspond to different answers to a single question: How does a given society produce the material goods needed for its reproduction over time? This question requires information concerning production techniques and the social control of the production process. First, what are the characteristic techniques of production available to the society? That is, what sorts of agricultural and finished-goods technologies are available? And second, what are the social relations of power and authority through which the production process is controlled? In other words, how are labor services allocated and controlled, how is the productive wealth of society controlled, how are finished goods allocated, how is the surplus allocated, and how is the technical process of production organized? Answers to the former question define the forces of production found within the given society, and answers to the second question define the relations of production.

In *The Communist Manifesto* Marx gives a fairly clear treatment of the forces of production. They are the instruments of production in use at a particular time, including "machinery, application of chemistry to industry and agriculture, steam navigation, railways, electric telegraph, clearing of whole continents for cultivation, canalization of rivers, whole populations conjured out of the ground" (*R1848*, p. 72). Thus they include the means by which society produces the goods necessary to satisfy the needs of its population: tools, labor power, raw materials, the technical division of labor, applied science, and so forth. They represent the level and form of technology present within a given productive system, irrespective of the social organization of production.

The relations of production include the social relations of domination and control through which the forces of production are utilized and developed, and through which the fruits of production are distributed. "The various stages of development in the division of labour are just so many different forms of ownership, i.e. the existing stage of the division of labour determines also the relations of individuals to one another with reference to the material, instrument, and product of labour" (*GI*, p. 43). These relations include the system of property and the system of the control of labor: ownership of means of production, tools, and labor power (capitalism); ownership of the worker (slavery); and control of the obligation to provide labor services (feudalism).

In a preliminary way it might be noted that the technical and social aspects of the production process are logically independent. We might imagine a variety of hypothetical societies that represent alternative assignments of forces and relations of production, for example, the technology of nineteenth-century industrial production paired with the social relations of primitive communal society. Historical materialism maintains, however, that only a minority of these hypothetical societies are in fact possible because existing forces of production limit the relations of production, and some sets of production relations are incompatible with certain forms of the productive forces. (For example, slave production relations are incompatible with a technologically sophisticated system of industrial production.)

Marx uses this taxonomy of the production process to formulate one of his most basic theses within the theory of historical materialism. This is his view that conflict between the forces and relations of production provides the dialectic that animates social change.[3] The forces of production are the raw power of society; they are regulated and exploited by the social relations of production. Marx argues, however, that as a given social system develops, the relations of production pass from enhancing the forces of production to limiting them, and once they have "become fetters on the forces of production," they must burst. Thus in discussing the transition from feudal production to capitalism Marx writes in *The Communist Manifesto*, "We see then: the means of production and of exchange . . . were generated in feudal society. At a certain stage in the development of these means of production and of exchange . . . the feudal relations of property became no longer compatible with the already developed productive forces; they became so many fetters. They had to be burst asunder; they were burst asunder" (*R1848*, p. 72). This is the basic hypothesis of *The Communist Manifesto*; it also appears in *The German Ideology* and *Capital*. The closing passage of *Capital* sounds the same theme, this time about capitalist production: "The centralization of the means of production and the socialization of labour reach a point at which they become incompatible with their capitalist integument. This integument is burst asunder. The knell of capitalist private property sounds. The expropriators are expropriated" (*Capital I*, p. 929). (See also the discussion of technological determinism later in this chapter.)

Let us turn now to Marx's concept of the economic structure of society. In the passage quoted earlier from the preface to *A Contribution to the Critique of Political Economy*, he describes the economic structure as "the relations of production appropriate to a given stage in the development of their material forces of production." Thus the economic structure is defined in terms of the social relations of production within which its productive activity is organized, for example, the property form, the form of labor organization, the form of surplus extraction, and the class system. Different modes of production are characterized by profoundly different relations of production, and these differences impose different laws of development and organization on the social formation. The relations of production of a given social formation impose a pattern of development on the most basic institutions of that social formation, and some modes of production lead by their own internal dynamic to the development of a new mode of production. The aim of Marx's political economy, therefore, is to discover and work out the specific relations of production that identify and distinguish the given mode of production from other modes, and to determine the laws of organization and development imposed by these relations on the mode.

The concept of the economic structure rests on the notion of a social relation of production. A social relation is an objective relation among men and women that is embodied in the structure of society through some set of incentives and penalties. It is stable over an extended period and is independent of the particular individuals participating in it, and it is typically reproduced through definite social institutions from one period to the next. The social institution of a library, for example, is a structure defined in terms of the stable relations among the persons involved in it: librarians, clerks, users, supervisors, and so on. The library is independent from the particular persons who occupy the roles in the sense that (1) the institution remains unaltered despite changes of personnel, and (2) disciplinary incentives exist to induce individuals to behave appropriately to their role. The same is true for the economic structure of a society. The positions defined by the economic structure are stable under changes of personnel; and sanctions exist against persons who fail to fill their roles appropriately – for example, within capitalism "insubordinate" workers lose their jobs. As an illustration, the institutions of classical slavery are defined by the distinct roles characterized by relations of domination and subordination that participants may occupy: slaveowner, overseer, gang boss, and slave. These positions are seen in terms of the powers and obligations which attach to the role; and those powers and obligations are institutionally rather than individually defined. Finally, the individual's conformity to the obligations of a role is enforced through effective social sanctions: economic and social incentives, in the case of the master, and coercive or punitive threats, in the case of the slave.[4]

These points hold true of social relations generally; but what is a social relation of production? A relation of production is one through which productive activity is organized. Not all such relations count as relations of production, however. For example, Cohen argues that work relations within the process of production are not production relations.[5] Work relations are relations corresponding to the technical division of labor within the production process, for example, the relations between sawyers using a two-man saw. These relations are defined by the character of the technology in use; they are not defined in terms of social powers possessed by some individuals over others. Rather, production relations are those relations of power and authority through which the wealth available to a society is used, controlled and enjoyed.[6] Paradigmatic examples of relations of production are the master-slave relation of classical slavery, the serf-lord relation of European feudalism, the wage-labor relation of modern capitalism, and the relations of authority found in factory production.

Thus the economic structure of society is defined by the social relations through which the productive process is controlled and directed, and through which the fruits of production are distributed. It is an objective structure in that it consists of relations that are independent of the wills of the participants. According to historical materialism, it is the "real foundation" on which rest the superstructure of society. The economic structure gives distinctive shape to noneconomic phenomena by imposing a set of constraints that limit the forms noneconomic institutions can take (what was referred to as a "logic of institutions" in chapter 1).

This account makes plain Marx's reason for selecting the economic structure of capitalism as the focus of his research in his economics: He expected that it would be possible to explain other fundamental aspects of capitalist society on the basis of knowledge about the economic structure. The economic structure is thus the concept that unifies much of Marx's scientific research. This concept allows one to summarize Marx's research program in his economics very concisely: Marx wanted to provide a theory of the economic structure of modern society that could serve as the basis for explanations of noneconomic phenomena. By providing a detailed analysis of the economic structure of capitalism, Marx believed that he could lay a foundation for an explanation of the most important structural and developmental characteristics of capitalist society.[7]

So far we have used the concept of the mode of production without explicit discussion. This omission is significant, since Marx uses this concept in several different ways without the precision characteristic of the other concepts of historical materialism. Most generally, he describes the mode of production as the material basis of social life. The mode of production embraces "the real process of production, starting out from the material production of life itself, and . . . the form of intercourse connected with this" (*GI*, p. 58). "This sum of productive forces, capital funds, and social forms of intercourse . . . is the real basis of what the philosophers have conceived as 'substance' and 'essence of man' " (*GI*, p. 59). However, Marx does not use the concept consistently. Sometimes the mode of production refers to the technical characteristics of the production process; sometimes it refers to the social characteristics of the process; and sometimes it refers to both social and technical aspects of the process (the relations and forces of production).[8]

Moreover, when Marx turns to serious analysis, he considers the economic structure rather than the mode of production in general. His chief concern—in *The German Ideology*, in *The Communist Manifesto*, and in *Capital*—is with the property relations and the relations of production found within a given social order. The mode of production is the concept that Marx uses most often in characterizing the dependency of noneconomic phenomena such as ideology or politics on facts about material life; the suggestion is that both technology and social institutions of production impose a "logic" on social life. But Marx's research in *Capital* is chiefly concerned with the social aspect of this thesis; he is not primarily concerned to discuss the implications of technology, but rather the implications of the social relations through which technology is utilized. And this fact in turn explains why the economic structure is the concept that receives his primary scientific attention. Thus the concept of the mode of production is subordinate to the concept of the economic structure in most of Marx's serious analysis of capitalism. This fact suggests that the economic structure is the central focus of Marx's research and that the mode of production is an umbrella term used in less rigorous contexts. It therefore seems reasonable to conclude that the concept of the mode of production plays a somewhat secondary role within the theory of historical materialism.[9]

The notion of the mode of production has at least one important methodological implication for social science conducted within the framework of historical materialism: the requirement of historical specificity. Since the mode of production is the structure of social organization through which human beings satisfy their material needs at a particular historical stage, there naturally will be certain elements in common among all modes of production. All involve organized labor; all use some form of tools and raw materials; all produce a material surplus; all involve natural scarcity; and all serve to satisfy human needs. So there is such a thing as production in general. "Production in general is an abstraction, but a rational abstraction in so far as it really brings out and fixes the common element and thus saves us repetition. . . . Nevertheless, just those things which determine their development, i.e. the elements which are not general and common, must be separated out from the determinations valid for production as such" (*Grundrisse*, p. 85). On Marx's view, then, modes of production differ crucially in the form of social relationships through which these ends are accomplished, and these differences impose radically different laws of development and organization on the different modes of production. Therefore the materialist must be concerned with the historically specific features of the mode of production—the social relations that distinguish the mode from others and that impose their own unique stamp upon the organization and development of the mode of production. The conclusion Marx draws from this observation is that the theory of the mode of production is a historically specific science: It is charged with producing a theory of a historically specific mode of production.

Relation of Economic Structure to the Social Formation

The object of Marx's inquiry is the economic structure of capitalism, which is distinct from the concrete social formation: the particular society as it exists at a given time and place.[10] What is the relation between these two levels of description? Briefly, the relation is that of an abstract pattern of social organization to the concrete society that embodies that pattern. Marx treats the economic structure as a structural characteristic of social formations that has an existence of its own, independent of the particular social formation (*Capital I*, pp. 96–98). And since the ultimate object of Marx's theory is capitalism in general, not the particular social formation presented by British society in the period he studied, it is necessary to discuss the implications of this distinction with some care.

Marx's view of the relation between the social formation (the particular, historically given social system) and the mode of production is displayed in a passage in the preface to *Capital*. There Marx writes,

> The physicist either observes physical phenomena where they occur in their most typical form and most free from disturbing influence, or, wherever possible, he makes experiments under conditions that assure the occurrence of the phenomenon in its normality. In this work I have to examine the capitalist mode of production, and the conditions of production and exchange corresponding to that mode. Up to the present time, their classic ground is England. That is the reason why England is used as the chief illustration in the development of my theoretical ideas. If however, the German reader shrugs his shoulder at the conditions of the English industrial and agricultural labourers, or in optimist fashion comforts himself with the thought that in Germany things are not nearly so bad, I must plainly tell him, *"De te fabula narratur!"* (*Capital I*, p. 90)

This passage offers three main conclusions. First, capitalism as a mode of production is subject to definite laws of development and organization (laws we reconstructed as expressing a "logic of institutions" in chapter 1). Second, the immediate object of Marx's investigation is British capitalism, that is, a particular social formation; but this is viewed as merely the most advanced manifestation of capitalism as a mode of production and not as its defining instance. And third, Marx believes that his analysis of capitalism applies to more than Britain alone; it provides knowledge about the development of German capitalism as well.

Thus Marx regards "capitalism" as being distinct from the British or German social formations. It is the abstract structure that these particular social formations share (perhaps imperfectly). Thus "capitalism" refers to the abstract structure that informs various social formations to a greater or lesser degree, the mode of production which underlying different social formations.

Moreover, Marx believes that the mode of production is fundamental in the task of explaining particular features of capitalist societies. This amounts to the claim that all capitalist systems develop along similar lines because they share the features identified in the theory of the capitalist mode of production (CMP). For this reason the theory of the CMP is explanatorily relevant to Japanese capitalism, even though Japanese capitalism did not exist when *Capital* was written. This belief depends on Marx's view that the institutions of capitalism impress a strong "institutional logic" on any society they inhabit—both economically (e.g., all capitalist systems show tendencies toward centralization of ownership) and superstructurally (e.g., all capitalist societies have an ideology supportive of private property). The former falls within the purview of *Capital*, and the latter falls within historical materialism proper.

Technological Determinism

Some interpreters maintain that historical materialism rests on what John McMurtry calls "technological determinism": the view that the forces of production determine the social relations of production (which in turn determine the non-economic structure).[11] This claim has provoked considerable disagreement among commentators. The issue concerns the question of the character of the "engine" of historical change. There are two broad families of possible factors. The propelling factor of change may be located either within the forces of production or within the relations of production. G. A. Cohen has provided the most extensive defense of the former position. He argues that for Marx, the forces of production are historically fundamental; as the forces of production develop they create relations of production that at first favor their development and then become obstacles to further development. When the forces and relations come into contradiction, the relations are "burst asunder" and a new set of relations of production appear that allow for the continuing growth of the forces of production. To make this thesis wholly intelligible, Cohen assumes that there is an inherent tendency for the forces of production to expand – a tendency he explains on the basis of facts about human nature (human needs expand, human reason is sufficiently powerful to see ways of improving current productive powers, and nature is generally not abundant enough to satisfy all human needs without labor).[12]

Cohen's argument depends heavily on Marx's statement of the theory of historical materialism contained in the preface to *A Contribution to the Critique of Political Economy*. In this formulation Marx describes the historical process as one in which relations of production first enhance, then hamper the development of the forces of production. When the relations of production become a fetter to the further development of the productive forces, they are "burst asunder." "Then begins an era of social revolution. The changes in the economic foundation lead sooner or later to the transformation of the whole immense superstructure" (*EW*, pp. 425–26). This formulation of the theory provides a good deal of support for Cohen's position, and similar formulations can be found throughout Marx's writings.

However (as Richard Miller argues at length),[13] the bulk of Marx's own historical explanations (in *Capital, The German Ideology, The Communist Manifesto*, and elsewhere) do not conform to the schema of technological determinism. When Marx attempts to explain the rise of the early bourgeoisie in "Primitive Accumulation" in *Capital* (*Capital I*, pp. 874–76, 914–26), when he attempts to explain the failures of proletarian revolution in 1848 in *The Class Struggles in France* (*SE*, pp. 45–47), or when he attempts to explain the creation of a "rabble of day-labourers" in the medieval town (*GI*, p. 70), his attention is primarily directed to the specifics of the relations of class and property within which these historical processes take place. "These workers [escaped serfs], entering separately, were never able to attain to any power, since, if their labour was of the guild type which had to be learned, the guild-masters bent them to their will and organised them according to their interest; or if their labour was not such as had to be learned, and therefore not of the guild type, they became day-labourers and never managed to organise, remaining an unorganised rabble" (*GI*, p. 70). The specific character of the existing forces of production does not enter into this explanation at all. These observations seem to indicate that Marx's own historical practice does *not* support technological determinism but rather the position that attributes explanatory primacy to the relations of production.

However, these arguments are not entirely conclusive, since they locate Marx's historical explanations at the level of relative class power and opportunity. It remains open to question, however, what factors determine the degree of power and opportunity available to diverse classes in society, and Cohen's view is that this question can best be answered by appeal to facts about the level of development of the productive forces. Cohen's version of technological determinism does not deny that the relations of production causally affect the productive forces; on the contrary, his functionalism entails that the reverse is true. What makes his position one of technological determinism is his view that the relations of production are selected because they lead to the enhancement of the productive forces.[14] From this point of view, the sorts of historical explanations cited earlier are interesting and important, but not fundamental, exactly because they do not identify the circumstances within the productive forces by virtue of which the decisive classes are decisive. John McMurtry's reconstruction of technological determinism has similar implications. On his view, the productive forces have explanatory primacy over the relations of production in that the former select out relations of production that do not comply with the productive forces;[15] but once again, it follows fairly immediately from this formulation that the relations of production have causal consequences on the productive forces.

It is not possible to resolve the issues surrounding technological determinism in this context. It is important, however, for us to arrive at a judgment about the role of technological determinism in *Capital*, and it is my view that *Capital* does not reflect such a position. Rather, *Capital* seems to attribute explanatory primacy to the relations of production that define the capitalist economic structure. For *Capital* seems to show that it is the social relations of production characteristic of capitalism that lead to its explosive expansion in technology and productivity, rather than expansion in technology leading to alteration in the social relations. For example, one of Marx's central arguments in *Capital* (discussed in chapter 3 below) is intended to explain the imperative toward the accumulation of capital that is entrenched in the CMP. It is this imperative which leads to the explosive increases in productivity and technology characteristic of capitalism. But this imperative follows *not* from some inherent tendency in the forces of production to expand but rather from the social relations of capitalism (in particular, the circumstances that both motivate and constrain the capitalist decision maker). Gary Young argues for this position convincingly in "The Fundamental Contradictions of Capitalist Production."[16] He maintains that the tendency of the forces of production to expand can be explained only on the basis of facts about the relations of production. Capitalism develops through the following contradiction: The relations of production both cause the forces of production to expand without limit and make it impossible to smoothly realize the price of the commodities so produced. This account therefore clearly locates the cause of the dynamic properties of capitalism within the economic structure, not within the forces of production themselves.[17]

Indeed, one of the features that most sharply distinguishes capitalism from other modes of production is the rapidity of the development of the productive forces. Classical slavery displays a fairly constant level of productivity, and feudalism witnesses a level of productivity which increases rather slowly over centuries. Capitalism on the other hand increases its level of productivity in leaps and bounds. And Marx's account of this difference depends entirely on his view of the different patterns of development imposed by the social relations which define these alternative modes of production.[18]

Finally, *Capital* focuses on the social relations that define the economic structure: the class system of capitalist and wagelaborer, the incentives and constraints that determine the behavior of the capitalist, and the theory of exploitation. Each of these has to do with the social relations that constitute the capitalist mode of production, not the productive forces.

These arguments suggest that technological determinism does not correctly characterize Marx's stance in *Capital*. However, even if this conclusion is accepted, the issue of the general theory of historical materialism is still unresolved. For throughout this work I will argue that it is necessary to distinguish clearly between the full theory of historical materialism and the specialized inquiry *Capital* represents. This distinction has consequences in the present context, for it is entirely possible that technological determinism does characterize historical materialism generally but does not characterize the approach taken in *Capital*.

Classes and Economic Structure

The concept of class plays a central role both in the theory of historical materialism and in Marx's theory of capitalism. I accept the view (argued by G. A. Cohen, Perry Anderson, and others)[19] that Marx has a structural theory of class: class membership is determined by one's position within the system of property. An individual belongs to the proletariat if and only if that individual (1) owns no productive wealth and (2) sells labor-power for a wage. This definition of class has two noteworthy features. First, it is an objective criterion of class membership, since it does not depend on class consciousness. (That is, one can be a member of the proletariat without identifying oneself as such.) Marx maintains, however, that objective class membership as defined here has explanatory import: Classes defined by the objective criterion tend to come to recognize their shared interests and thus to constitute self-conscious units of political activity—that is, classes tend to become class-conscious. Thus the notion of a "class-in-itself" is explanatorily prior to that of a "class-for-itself," although the latter concept is central to explanations of class political activity and revolutionary motivation.[20] And second, this definition of class ensures that classes will have conflicting material interests, since their positions are defined by an exploitative property system.

It is noteworthy that *Capital* uses only the structural criterion of class; there is little developed analysis of the factors that tend to make a class into a politically unified organization. The only explicit discussion of classes in *Capital* occurs in the final chapter of volume III, and this discussion is notoriously fragmentary (*Capital III*, pp. 1025–26). The concept of class occurs throughout the work, however, but almost always in the form of objective class identity rather than subjective class consciousness. Thus Marx writes in the preface to *Capital* that "individuals are dealt with here only in so far as they are the personifications of economic categories, the bearers of particular class-relations and interests" (*Capital I*, p. 92). This comment plainly makes the concept of class explanatorily central, and equally plainly, it is the structural definition of class that is at work here—class as the embodiment of an economic category. (A satirical example of this sort of personification occurs at the end of part two of *Capital*: "He who was previously the money-owner now strides out in front as a capitalist; the possessor of labour-power follows as his worker. The one smirks self-importantly and is intent on business; the other is timid and holds back, like someone who as brought his own hide to market and now has nothing else to expect but—a tanning"; *Capital I*, pp. 279–80.) An important exception to the structural definition of class in *Capital* is found in the penultimate section, where Marx describes the "revolt of the working class, a class constantly increasing in numbers, and trained, united and organized by the very mechanism of the capitalist process of production" (*Capital I*, p. 929). However, when in *Capital* Marx examines in detail the logic of the capitalist mode of production, it is the structural conception that is at work, with its associated notions of interest and incentive motivating the individual.

Thus for the purposes of analysis in *Capital* the theory of class can be reduced to two simple points: The economic structure is defined by a particular form of property, and separate groups in society fall into separate locations within that structure. The sparsity of this account follows from the fact that *Capital* is an economic rather than a political analysis. And it reinforces the conclusion that

Capital is fundamentally a study of the economic structure of capitalism and the logic it imposes on the mode of production and not an analysis of the characteristics of the superstructure—politics or class consciousness.

An important issue in this context is the question of how the needs of the economic structure are translated into actual social change. The theory of class provides at least a partial solution. G. A. Cohen argues that class struggle is (often) the means by which the economic structure influences other aspects of social structure.[21] A class consists of a group of persons with a set of interests determined by their position within the economic structure. Insofar as these persons pursue their interests rationally, they also serve the "needs" of the elements of the economic structure which they represent. The economic structure "needs" a set of political institutions that support existing property relations; and the class that benefits from those property relations is in a position to influence state actions in that direction. In order to explain why a given class is dominant at a certain time, it is necessary to refer to the economic structure and to the forms of power that membership in the economically dominant class makes available to that class. The power of the bourgeoisie in the period of decline of the precapitalist economy in England is to be explained by the capacity of members of that class to control of a set of forces of production that were dramatically more productive than those associated with traditional forms of property (i.e., traditional landownership). Thus it is through the rational activities of members of a class that the economic structure imposes its imprint on other social institutions.

Primacy and Determination

Determination by Constraint

Historical materialism maintains that the economic structure has explanatory *primacy* over noneconomic structure. This relation between different levels of social structure depends on the idea that one level "determines" the other. This notion requires further consideration, however. Through what sort of relation can one level of social organization be said to explain the other? Strict causal determination of higher-level structures by lower-level factors is not the relation at work. There is plainly a good deal of latitude between the economic structure and the superstructure—witness the variety of political forms found among capitalist economic systems (e.g., representative democracies, constitutional monarchies, or military dictatorships). So determination in this context does not entail *unique* determination. Most Marxian interpreters agree that the relation is not one of strict causal determination at all. The wage relation does not mechanistically *cause* the development of the ideology of fair contracts.[22]

Two related models of determination have emerged in recent years that permit a nonmechanistic interpretation of the notion of determination. One of these is John McMurtry's notion of determination by constraint. This views the relation between base and superstructure as a limiting one: "The economic structure determines the legal and political superstructures, the ideology, and the forms of social consciousness by blocking or selecting out all such phenomena that do not comply with it."[23] Thus superstructures tend to correspond to their economic bases because the base blocks the emergence of superstructural features incompatible with the continued working of the base. Similarly—for those who hold technological determinism to be true—economic structures tend to correspond to existing productive forces because these forces block the emergence of relations of production that are not compatible with them. On this sort of account, the relation between base and superstructure is one in which the base imposes constraints on the forms the superstructure can take. Further, economic institutions create a suitable environment for their reproduction and development through a negative process of filtering out superstructural institutions inconsistent with their requirements. Only certain structures are compatible with a given mode of production; if a structure incompatible with capitalism should take hold, one or the other must disappear. And on Marx's view, the structures associated with the system of production are particularly securely established, with mechanisms for repressing opposing tendencies; in general, then, the mode of production will exercise control over higher-level structures by weeding out incompatible ones.

This view of the relation between various levels of structure within society is thus an indirect one: The economic structure does not *directly* determine the noneconomic, but it imposes constraints on those structures that over the long run establish a strong tendency to develop toward a particular form of social organization. Certain structures cohere much more successfully with the capitalist substructure, and so it is likely that these will proliferate.

In order to understand fully this view of the relation between the economic structure and noneconomic phenomena, it is necessary to describe the mechanisms through which the lower-level structures constrain or filter superstructural elements. The filtering may occur through a variety of mechanisms, both intended and unintended. First, the needs of the economic structure are to a substantial degree reflected by the needs of the class corresponding to the dominant relations of production within the economic structure; and to greater or lesser degrees, classes may be expected to take rational action to preserve their interests and satisfy their needs. For example, the capitalist economic structure "needs" to accumulate capital. Correspondingly, the capitalist class needs to be free to earn profits on private property and has the means (in the form of economic and political power) to protect and preserve this power. By acting to protect their freedom to earn profits, however, members of the capitalist class also serve the need of the economic structure to accumulate capital. Consider an example of such a result: A political party within a capitalist democracy that promised to nationalize the steel industry could not expect to receive financial support from the capitalist class; but without massive financial backing, parties within a capitalist democracy cannot expect to be successful, since access to the media (newspapers, radio, television, printers, etc.) is gained largely through advertising budgets.

A second mechanism by which the economic structure filters superstructural elements is unintentional; that is, it does not result from the deliberate actions taken by classes to protect their position. This sort of filtering stems from the conditions imposed on society at large by the requirements of a smoothly functioning capitalist economy, chief among which is that of profitability in capitalist enterprises. Suppose, for example, that the Commonwealth of Massachusetts enacts a stringent new set of health and safety requirements in industry that are costly to individual capitalists, and substantially more so than those imposed on industry in New Hampshire. As individual capitalists work out the probable rate of profit in Massachusetts under the new conditions versus that in New Hampshire, plant closings result in Massachusetts. Tax revenues fall in Massachusetts, unemployment in the state rises, and pressure mounts (from unemployed workers, city officials, municipal workers, small business owners whose incomes depend on sales to industrial workers, etc.) for revision of the requirements. These pressures lead ultimately to reform of the new laws, and new requirements are enacted that are more compatible with the requirements of profitability. In this case a superstructural innovation (the new health and safety laws) is put into place that is incompatible with the capitalist economic structure (in particular, the requirement that capitalists act so as to maximize the rate of profit), and feedback mechanisms arise that compel revision of the new laws.

An example of filtering that contains elements of both the aforementioned forms concerns the general tenor and biases of the ideological institutions of capitalist democracies. The ideological system is the set of institutions that influence and shape the attitudes, beliefs, and values commonly held within a given society about human nature and society; it is Marx's view that this system is by and large conditioned by the needs of the system of production. How is this conditioning thought to obtain? We might first note that social consciousness is shaped largely by institutions like newspapers, television, schools, books, and family life. Most of these, however, are influenced by class considerations, and points of view genuinely dangerous to the existing class structure can be effectively filtered out. Newspapers, publishers, and educators commonly reflect the dominant ideology that infuses their products; and it is these products that are responsible for shaping public opinion and attitudes. "Radical" points of view—those that seriously undermine or challenge the existing social system—are "irresponsible" and are denied access to these means of shaping public opinion. (Marx's version of this process appears in the preface of *Capital* with respect to the condition of political economy in Germany; *Capital I*, pp. 96–97.) This view does not require a determination of thought by the underlying economic structure; it does not imply a completely relativistic theory of knowledge. It only emphasizes the powerful institutional forces that bias social consciousness in the direction required by the system of production.[24] The point is not that radical points of view are totally suppressed or that it is impossible to publish such views; it is rather that such viewpoints must confront pressures that make their eventual appearance uncommon. They are generally drowned out by proponents of the prevailing ideology. In this example the economic structure conditions higher-level social structures by being securely established and ruling out (or at least making very unlikely) those structures that contradict it or interfere with its functioning.

Functional Determination

The other prominent model of primacy is G. A. Cohen's conception of functional determination. On Cohen's view, the superstructure possesses its particular characteristics by virtue of the functional requirements of the base that those characteristics satisfy; likewise, the economic structure may be explained by reference to the functional requirements of the productive forces. "We hold that the character of the forces [of production] functionally explains the character of the relations [of production]. . . . The favoured explanations take this form: *the production relations are of kind R at time t because relations of kind R are suitable to the use and development of the productive forces at t, given the level of development of the latter at t.*"[25] And again: "When relations endure stably, they do so because they promote the development of the forces."[26] This sort of account may be described as a "consequence explanation" because it is an explanation in which "something is explained by its propensity to have a certain kind of effect."[27] On this account, the relation between economic structure and superstructure is a functional relation, not a direct causal one. Among the superstructural features that would be compatible with the base, some are more suitable to the ongoing needs of the system. And the functionalist thesis maintains that (1) when political or ideological elements appear that are inconsistent with the continuing survival of the economic structure, they will usually be weeded out, and (2) when competing superstructural elements appear that are each compatible with but differentially suitable to the economic structure, the more suitable will generally survive rather than the less suitable. The resulting social system will be one in which upper-level structures are largely suited to the functional needs of the underlying level of organization (in the short run), and it will seem as though the underlying system directly shaped the upper-level structures — although there is in fact no direct causal relationship between the two.

Cohen acknowledges that functional explanations may be supplemented by an account of the processes by which superstructural features come to have the properties "needed" by the base. He calls such accounts "elaborations" of the functional explanation. These elaborations are largely similar to the varieties of filtering processes described earlier.[28] However, Cohen insists that the correctness of a functional explanation does not depend on the availability of an acceptable elaboration of the explanation; rather, the functional explanation can be confirmed empirically.[29]

Cohen's functionalism has been subject to extensive criticism. In an important treatment of functionalism, Jon Elster argues that "[in functionalism] Marxist social analysis has acquired an apparently powerful theory that in fact encourages lazy and frictionless thinking."[30] His view is that functionalist explanation has no proper role in social science because there is no appropriate account available of how social institutions could come to acquire functional characteristics — either by design in the case of artifacts or through natural selection in the case of biological systems.[31] Consequently Elster maintains that functionalist explanations in social science are vacuous unless supplemented by appropriate "microfoundations" — detailed analysis of the processes by which given social patterns emerge.[32] And once such microfoundations are available, the functional explanation is no longer needed. John Roemer supports this position: "I argue that class analysis requires microfoundations at the level of the individual to explain why and when classes are the relevant unit of analysis."[33]

This issue is somewhat cloudy, since it is not entirely clear that the opposing positions are genuinely incompatible. Cohen contends that historical materialism is primarily concerned to establish macrolevel functional relations between large units of social structure, whereas Elster holds that historical materialism should be primarily concerned with discovering the microlevel processes by which such relations emerge. But Cohen does not deny that there are such processes or that it is an important problem of social science to work them out in detail, and Elster's position does not entail that functional relations do not exist between levels of structure. So this issue seems ultimately to come down to a difference of emphasis within a program of research, rather than a substantive disagreement over the nature of social change.

Functional determination is more vulnerable to attack from another direction. This vulnerability concerns its requirement that every aspect of the economic structure serve particular functional needs of the productive forces (and likewise for the relation between superstructure and economic structure). This is a substantially stronger form of determination than that postulated by the constraint model. It is in fact too strong: Given the sorts of elaborations Cohen himself puts forward, and the sorts of social processes by which lower-level factors influence higher-level ones, it is perfectly possible that higher-level institutions will emerge that neither promote nor hinder lower-level institutions. By contrast, the constraint model requires only that superstructural features be *compatible* with the base, but particular features of the superstructure may be neutral with respect to the functional needs of the base. There will generally be a range of possible superstructural institutions compatible with a given form of the economic structure. And it is perfectly consistent with the constraint model that the factor that determines *which* of these possibilities will emerge may be noneconomic considerations. (For example, the fact that Italian society is overwhelmingly Catholic may determine which of the possible state forms will in fact come into being.) In this aspect the constraint view seems more convincing than the functional determination view.

Moreover, the constraint view permits us to capture the most plausible portions of the functionalist account, since the constraining process will frequently give rise to institutions functionally suited to the economic structure of the given society. And where the two models differ — in that the constraint model is consistent with superstructural elements that do not serve functional needs of the base — the constraint model seems more plausible. I conclude, therefore, that the most convincing reconstruction of the notion of correspondence is the constraint view, according to which the economic structure limits the range of possible political and ideological forms, without strictly ruling out a great variety of forms. The economic structure imposes broad limits on the characteristics that institutions must have, and this forces certain functional and structural characteristics on social institutions.

Both these accounts succeed in obviating a classical objection to historical materialism: the charge that historical materialism requires that lower-level factors (productive forces or relations of production) determine the precise character of upper-level factors (the elements of the superstructure). For both accounts leave room for a substantial degree of autonomy for noneconomic institutions with respect to the economic structure. The constraint model makes it apparent that a wide range of superstructural institutions will be compatible with the base, since the compatibility requirement is a fairly weak form of correspondence. This dimension of variability at the level of the superstructure is inherent in the notion of a functional relationship as well. It is *possible* for a desert rodent to have water-wasteful habits — although that would require compensatingly effective mechanisms for gathering water in a dry environment. Similarly it is possible for a capitalist society to contain ideological elements that are not best suited to its growth or to exist within an anachronistic state form. These anomalies are possible; but they are improbable, and to the degree they hinder the full development of the economic structure, they are unstable. There may well be a range of possibilities at a given time that are more or less equally suitable to the given economic structure, however, or even a range of possibilities which are differentially suitable but nonetheless feasible; for instance, capitalism may be less productive under an absolutist state but still economically viable. Thus on either model it is perfectly plausible to hold that there will be a variety of different state forms associated with the capitalist economic structure.

Given this substantial degree of independence possible for noneconomic institutions, a full theory of capitalism unavoidably requires theories of politics and ideology, and these theories are not derivable from a theory of the economic theory. Some of Marx's successors (e.g., Gramsci)[34] have developed this insight in the direction of providing a fuller theory of politics than is entailed by the thesis that the bourgeois state is the managing committee of the ruling class. Ralph Miliband's *The State in Capitalist Society*[35] and Nicos Poulantzas's *Political Power and Social Class* are examples of contemporary work inspired by the same insight. And this fact goes a long way toward satisfying E. P. Thompson's criticisms of *Capital* in *The Poverty of Theory* (discussed in chapter 7), since it establishes that the theory of the economic structure—even if complete—could not in principle determine the correct theory of politics.

Historical Materialism and *Capital*

In spite of the extensive attention devoted to historical materialism in recent years, the question of the relation between historical materialism and *Capital* has not been sufficiently considered. Most writers tend to presuppose what may be called the subsumption theory. This view holds that historical materialism is a general theory of history and that *Capital* is the specialized application of that theory to capitalist society. According to this position, *Capital* is entirely contained within the framework of historical materialism in that its concepts, methods, and research goals are defined by the earlier work. *Capital* is the application of these ideas to the particular features of capitalist society.

This subsumption theory is shared by a wide range of commentators on Marx. Thus Nicos Poulantzas writes, "Historical materialism maintains a *general* theory defining the concepts which command its whole field of investigation (the concepts of mode of production, of social formation, of real appropriation and property, of combination, ideology, politics, conjuncture and transition). . . . Historical materialism also includes *particular theories* (theories of the slave, feudal, capitalist and other modes of production)."[36] And according to Poulantzas, *Capital* is just such a specialized theory. In a similar vein I. I. Rubin writes, "There is a tight conceptual relationship between Marx's economic theory and his . . . theory of historical materialism. . . . Theoretical political economy deals with a definite social-economic formation, specifically with commodity-capitalist economy. . . . Marx's theory of historical materialism and his economic theory revolve around one and the same basic problem: the relationship between productive forces and production relations. . . . By applying this general methodological approach to commodity-capitalist society we obtain Marx's economic theory."[37] And Maurice Godelier describes *Capital* in similar terms: "The method of *Capital* is formed on the basis of the philosophical assumption of materialism. This philosophy is enveloped in the heart of the theory which it has made it possible to develop. *Capital* therefore presupposes the critical movement that led Marx to dialectical idealism and then to materialism through the *1844 Manuscripts, The German Ideology*, etc."[38]

The subsumption theory has an equally strong hold on Anglo-American commentators on Marx. Thus John McMurtry assimilates historical materialism and *Capital* by suggesting that both investigations are concerned with uncovering the "laws of motion" of various modes of production: "In the case of capitalism, these 'laws of motion' are identified in extended detail; . . . unlike his analysis of the 'motion' of other economic structures, [Marx] formalizes and schematizes this whole process. . . . With the particular economic structure of capitalism, in short, the 'laws of motion' governing the ruling-class pattern of production relations constitute a massive theoretical system."[39] Here the difference between historical materialism and *Capital* is a matter of degree of detail—not a difference in method or scope. G. A. Cohen appears to presuppose the subsumption view as well. He regards historical materialism as a general theory of history: "We may attribute to Marx . . . not only a philosophy of history, but also what deserves to be called a *theory* of history."[40] And his free use of *Capital* in developing his reading of Marx's theory of history implies that he regards *Capital* as part of that theory.[41] This reliance on *Capital* suggests that Cohen sees no sharp distinction between historical materialism and Marx's economics; rather, *Capital* is one of the chief works in which Marx's theory of history is worked out and applied.

In spite of widespread agreement on the subsumption theory, this view misses the mark. The scientific content of *Capital* is substantially distinct from historical materialism, and the subsumption theory conceals important differences of scope and rigor between these areas of Marx's thought. This point has several bases. First, the scope of *Capital* is substantially narrower than that of historical materialism. *Capital* is a limited and specialized study of capitalist society. It is concerned only with the capitalist *economic* structure (not its political, cultural, or ideological aspects). As a result *Capital* is not committed to many of the most interesting but controversial claims of historical materialism, for example, that the economic structure determines politics or ideology. Second, Marx uses empirical evidence in *Capital* much more extensively and precisely than he does in his chief writings on historical materialism. And finally, the two areas of Marx's thought differ in their standing as works of science. Historical materialism is a general set of ideas that make up at most a program of research, whereas *Capital* is a specific and developed empirical analysis.

The chief harm in the subsumption theory is that it obscures the scientific merits of Marx's economic analysis by assimilating it to the less rigorous hypotheses of historical materialism. *Capital* goes significantly beyond Marx's earlier achievements in its standing as a work of social science. By insisting on the important distinctions between these two areas of Marx's thought, it is possible to evaluate each on its own merits.[42]

Before turning to the arguments needed to establish these conclusions, it must be acknowledged that there is an important ancestral relation between historical materialism and *Capital*. Historical materialism provided Marx with the general problem of research for *Capital* by concentrating his attention on the social relations of production that constituted the economic structure of capitalism. Historical materialism maintained that a stringent analysis of the economic structure of society was needed to explain other social phenomena, and *Capital* was intended to provide that analysis.[43] Further, historical materialism provided some of the general concepts Marx used to define his project (economic structure, production relations, mode of production, class, property, and the historicity of social relations). These concepts function as a general background ontology in *Capital*, and they led Marx to define his investigation substantially differently from the classical political economists in that he emphasized the historical specificity of the social relations which characterized capitalist production.[44] Historical materialism is therefore important to understanding the research goals and presuppositions Marx brings to his economics. But having established the importance of research into the economic structure, historical materialism is largely silent on the substantive and methodological problems of conducting that investigation.

A second qualification must be made as well. In rejecting the subsumption theory, I do not hold that *Capital* overturns the major findings of historical materialism or that there are major inconsistencies between the two areas of Marx's thought. Rather, I hold only that *Capital* is a substantial advance beyond historical materialism: It uses different methods and concepts to investigate more specialized questions, and it is not committed to most of the substantial claims of historical materialism. It therefore cannot be subsumed under the theory contained in the chief writings of historical materialism.[45]

Scope and Limits of *Capital*

Let us first consider differences in the scope of the two investigations. Historical materialism is a general hypothesis about the dynamics of social change and organization in all societies. It maintains that facts about the economic structure and technology—the "material foundation"—of any society are fundamental to explanations of noneconomic institutions. The specific character of the forces and relations of production found within a given society, and the contradictions that eventually develop between them, are said to impose a logic on the noneconomic institutions of the society. The result of this postulated primacy for economic factors is that historical explanation must be founded on analysis of these forces and relations of production. Historical materialism is thus a *general* theory in two respects: It is a theory of many different modes of production, and for each mode of production, it is a theory of the full social system – not merely the economic structure of the system.[46]

By contrast, Marx's research in *Capital* is doubly restricted. *Capital* looks at only one mode of production, and within that mode it is concerned only with the logic of the economic structure, not the whole of capitalist society. Marx's aim in *Capital* is to uncover the "economic law of motion of modern society" (*Capital I*, p. 90). These laws of motion are the long-term tendencies of development observable in the capitalist economy. They include the falling tendency in the rate of profit, the recurring crises of capitalism, the creation of an industrial reserve army, concentration and centralization of production, and continuing class separation in capitalist society. Marx regards these as empirically observable tendencies, and he seeks to provide an abstract model of the economic structure of capitalism that would account for them.

Thus Marx's research problem in *Capital* is quite narrow: to investigate the institutional logic created by the economic structure of capitalism. Capitalist society has many different aspects – cultural, ideological, political, and so forth. Marx's purpose in *Capital* is limited: to discover the laws of the capitalist economy and to provide an account of the logic of the supporting economic institutions. For in the same passages in which Marx speaks of the laws of motion of capitalist society, he emphasizes that these are to be *economic* laws of motion. In posing this research problem, Marx thus puts aside the problem of capitalist society in general in order to focus on the economic system. Thus – unlike historical materialism – *Capital* is not intended to offer a complete theory of capitalist society as a whole.

The limits Marx imposes on the scope of his account are apparent both in his express declarations about his purposes and in the content of *Capital* itself. Throughout his many plans of research from the 1850s onward, Marx consistently describes his economics as merely the first installment of a more extensive analysis of capitalist society. The economics was to focus on the economic institutions of capitalism, and a later volume was to consider the political characteristics of capitalist society.[47] Whenever Marx describes his purposes in *Capital*, he does so in a way that reflects this definition of subject matter: "What I have to examine in this work is the capitalist mode of production, and the relations of production and forms of intercourse that correspond to it"; "It is the ultimate aim of this work to reveal the economic law of motion of modern society"; "My standpoint [is one] from which the development of the economic formation of society is viewed as a process of natural history" (*Capital I*, pp. 91, 92, 93; emphasis added).

A survey of the main concerns of *Capital* confirms this view. Marx's research problems derive from classical political economy, not primarily from the chief writings of historical materialism. What causes chronic unemployment? What leads capitalist industry to technological innovation? What factors influence the long-term behavior of the rate of profit? What is the source of profit, interest, and rent? These questions make use of the specialized conceptual scheme of political economy (value, price, profit, accumulation, etc.) rather than the more general concepts of historical materialism (class, ideology, mode of production). These questions are comparatively narrow in scope and specific enough to admit of scientific treatment.

Further, the scientific content of *Capital* is largely independent of the chief concerns of historical materialism—the theories of politics and ideology. Marx does not discuss these areas extensively in *Capital*, and his economic arguments rarely turn on substantive assumptions about ideology and politics. In "The Working Day" Marx refers to the political power of the working class and discusses the Factory Acts, and in "So-Called Primitive Accumulation" he discusses the Enclosure Acts (*Capital I*, pp. 375–411, 876–904). But neither context represents a substantive contribution to political theory, nor is it Marx's intention to do so. Furthermore, though Marx no doubt believed that the capitalist economy required various forms of state support and intervention in order to function smoothly, he does not take it upon himself to offer a rigorous or extensive analysis of the ways in which state power could be marshalled by the possessing classes in defense of their interests.

Marx's discussion of ideology in *Capital* is equally unsystematic. In chapter 1 he discusses the "fetishism of commodities" and offers an account of the origins of certain pervasive misconceptions held by the participants in a capitalist economy. But these comments function as suggestive asides rather than sustained analysis, and they are theoretically superfluous because they do not serve as the basis for other arguments elsewhere in *Capital*. Moreover, the view of social consciousness advanced here is to some degree inconsistent with the theory of ideology presented in *The German Ideology* and *The Communist Manifesto*, since it does not emphasize the link between class power and consciousness that characterizes the latter works. Elsewhere in *Capital* Marx speaks of the ideology of exchange ("the very Eden of the innate rights of man"; *Capital I*, p. 280), and of the influence of ideology on scientific political economy (*Capital I*, pp. 96–98), but again his remarks are unsystematic and familiar. There is little progress here over similar assertions in *The German Ideology* or *The Communist Manifesto*.

These points show that Marx did not devote significant attention to the problems of politics and ideology in his research in *Capital*. *Capital* does not present developed theories of politics or ideology, and its central arguments do not depend on assumptions about politics or ideology derived from the theory of historical materialism more generally. Moreover, it was reasonable for Marx to disregard these areas. This is not because an economic theory of capitalism suffices to replace all other investigations, but rather because scientific research depends on limiting the scope of investigation. Thus Marx could hold that capitalist society relies on irreducible elements of noneconomic structure (e.g. political institutions) and even that the economic structure itself requires these institutions (e.g., in the form of legal guarantees of property rights), but still confine his attention to the strictly economic process of the system. Marx properly chose to leave noneconomic phenomena to other specialists working within the general framework of historical materialism.

Thus the contents of *Capital* bear out the claim that the scope of that work is strictly limited. Marx was not considering capitalism as a whole, but rather its economic structure in isolation from noneconomic institutions. Having identified the economic structure as being explanatorily primary, he turned away from the full generality of historical materialism to a more specialized discipline, political economy. In sum, *Capital* is a detailed analysis of the economic structure of capitalism and the dynamic of development imposed by that structure on the economic system. Its scope is narrow and it is largely independent of the more general claims of historical materialism. *Capital* therefore has at least this feature in common with paradigm works of science: it offers a specialized but restricted analysis of a carefully bounded range of phenomena.

Evidence in *Capital*

Let us now consider a second important difference between historical materialism and *Capital*: Marx's use of empirical evidence. Historical materialism postulates the need for empirical investigation: "The premises from which we begin are not arbitrary ones, not dogmas, but real premises from which abstraction can only be made in the imagination. . . . These premises can thus be verified in a purely empirical way" (*GI*, p. 42). Paradoxically, however, the chief texts in which historical materialism is developed do not contain extensive empirical detail.[48] Neither *The German Ideology* nor *The Communist Manifesto* provides an exemplary model of empirical or historical investigation. These works touch lightly on the history of European society, but what emerges is an interpretive overview rather than a careful historical analysis grounded on empirical evidence. Consider, for example, Marx and Engels's survey of the series of property forms found in European history. This discussion outlines a model of explanation: that social formations may be distinguished on the basis of the form of ownership they embody; and it illustrates that model with a sketchy discussion of ancient communal property, classical slavery, and feudal property. But Marx and Engels do not consider the historical evidence in the quantity and with the precision necessary to give their account standing as a serious historical treatment.[49]

By contrast, Marx's study of capitalism was based on a rigorous treatment of available evidence.[50] He uses historical and empirical matter in a variety of ways in *Capital*: as examples of theoretical points, as documentation of notably egregious features of capitalism, as support for particular theoretical assertions, and as a means of evaluating the truth of an abstract theory. Thus Marx's empirical reasoning is more complex than simple models of confirmation suggest. But it is unmistakable that Marx was committed to an empirical evaluation of his theory of capitalism. His theoretical research was constantly informed by wide reading in the economic history of capitalism, the contemporary behavior of English capitalism (e.g., fluctuations in prices and swings in the business cycle), and the social conditions current within English capitalism (the Blue Books, newspaper accounts of urban conditions, etc.).[51] Marx's abstract analysis of the capitalist mode of production thus rested on an integration of evidence drawn from the European history, sociological descriptions, and accounts of its present organization drawn from political economy.

Capital reflects this disciplined attention to empirical evidence. Chapter 25 ("The General Law of Capitalist Accumulation") offers a fair example of Marx's use of data. Here Marx works out some of the chief implications of the theory of capital for the observable behavior of the capitalist economy, particularly the tendencies toward a permanent "surplus population" within the work force (the industrial reserve army) and toward concentration and centralization. He then uses statistics on the English economy to argue for the correctness of the theoretical analysis.[52] Later volumes of *Capital* show a similar practice. In volume II Marx discusses the logic of reproduction and then considers economic data relevant to assessing his analysis; and in volume III he examines the dynamic tendencies of the rate of profit and evaluates this analysis in terms of the actual behavior of the rate of profit.[53] In each case we find a disciplined effort to rely on hard evidence to assess the theoretical principles.

These arguments show that Marx makes substantial and precise use of empirical data in *Capital*. He recognized that the ultimate criterion of his theory's success was the degree to which it was supported by the available empirical data concerning the capitalist economy, and he made rigorous use such data. Marx's use of evidence in *Capital* thus stands in sharp contrast to that in *The German Ideology*: *Capital* represents the sort of consideration of empirical data demanded by scientific inquiry, whereas *The German Ideology* does not. (Marx's empirical practices in *Capital* will be considered more fully in chapters 6 and 7.)

Historical Materialism as a Program of Research

These considerations suggest that historical materialism and *Capital* are not comparable works of scientific research. Instead, historical materialism is best construed as a research program for the social sciences, whereas *Capital* is a developed scientific analysis of limited aspects of the capitalist mode of production that falls within that program. Historical materialism provides a hypothesis about the process of historical change generally: that the economic structure "determines" noneconomic structure. And it embraces a variety of more specific hypotheses as well — theories of class, politics, ideology, religion, and so forth. But none of these is a theory in the full sense relevant to science, and consequently Marx has no scientific theory of history.

On this account, historical materialism is best seen as a family of research hypotheses that may serve as the basis for detailed empirical investigation; it is a "leading thread" or "guiding principle," as Marx puts it in the preface to *A Contribution to the Critique of Political Economy*, for more specialized forms of research. (*EW*, p. 426) The notion of a research hypothesis has played an important role in recent philosophy of science. Generally speaking, a research hypothesis is a notion which a scientist regards as worth pursuing without yet having been formulated fully enough to admit of empirical evaluation. Thomas Kuhn, Imre Lakatos, Larry Laudan, and others have pointed out the centrality of such ideas in scientific research;[54] they include Benjamin Franklin's notion that "electricity is a fluid," the eighteenth-century physicist's idea that "light is a wave," and Marx's contention that "history is a history of class struggle." Each represents a fruitful beginning for empirical and theoretical research; each provides an initial hypothesis in terms of which to organize and analyze data. Ideas at this level are of undeniable importance to the conduct of science, but they are not full-fledged theories. They are not sufficiently elaborated to constitute explanations, and they are not precise enough to have definite empirical consequences. They represent the beginnings of scientific knowledge rather than its end point.

The theses of historical materialism are just such research hypotheses. Historical materialism offers a general hypothesis about the causes of large-scale social change, and it invites social scientists to direct their research in lines suggested by that hypothesis. This function of historical materialism is scientifically important, but it is not the same as that fulfilled by a fully developed scientific theory. In particular, historical materialism cannot be said to have definite empirical content until more specific research has been completed under its direction.[55]

This view of historical materialism treats it as a portion of a research program — part of a method of investigation. But even this conclusion requires qualification. For even as a methodology, historical materialism is only of limited use to Marx's research in *Capital*. This limitation is inherent in the definition of historical materialism, since its chief methodological import amounts to the dictum "Seek out the economic structure." But precisely because this injunction takes the economic structure as explanatorily fundamental, historical materialism does not offer the basis for an analysis of the constituent parts of the economic structure itself. Historical materialism is concerned with the relations between the economic structure and the superstructure, whereas *Capital* is concerned with the internal logic of the economic structure. In order to provide such an analysis, it is necessary to formulate a more fine-grained method and theory than historical materialism can provide.

Significantly enough, Marx's methodological development did not stop with his writings of the late 1840s (the peak of his thinking on historical materialism). The introduction to the *Grundrisse*, *Theories of Surplus Value*, and portions of *Capital* further develop his conception of scientific method. The method that emerges from these works is firmly within the general perspective of historical materialism, but it offers more specific direction in explaining the organization and development of the economic structure of society. This theory of science consists of a conception of the nature of scientific explanation, a developed view of the form of abstraction appropriate to social science, and a full appreciation of the role of evidence in evaluating scientific claims.[56] Thus Marx went significantly beyond historical materialism by constructing a methodology for social science that guided his economic research.

Both historical materialism and *Capital* are major contributions, and they are plainly connected. However, it must be emphasized that they are distinct systems of thought. Historical materialism is not a full-fledged work of science. It is a fruitful program of research that suggests a model of explanation applicable to a wide variety of different historical phenomena, but it is not a developed empirical hypothesis. By contrast, *Capital* is specific and detailed in ways in which historical materialism is not; it more narrowly limits its claims; and it uses evidence in a more extensive and disciplined way. *Capital* thus goes beyond historical materialism by providing an empirically specific account of the capitalist economic structure, an account that is a clear contribution to social science.

3

Marx's Economic Analysis

In chapter 1 we argued that Marx's account of capitalism is not a unified deductive system along the lines of theories in natural science. Instead, *Capital* presents a selective description of the defining structural and functional features of the capitalist economy and an analysis of the "logic of institutions" to which those features give rise. Chapter 1 also surveyed the main elements of Marx's descriptive account. This institutional-logic treatment gives a sociological interpretation of *Capital* in that it emphasizes the centrality of Marx's depiction of the fundamental social relations of production of capitalism. This approach is correct as far as it goes, but it must be supplemented by a treatment of the economic reasoning in *Capital*. For whatever else it is, *Capital* is a work of economic analysis. It is through his reasoned answers to a family of technical economic questions about the capitalist mode of production that Marx arrives at the "laws of motion" of that economic form. This chapter will develop this treatment of *Capital* by considering the nature of Marx's economic reasoning: the forms of economic argument by which he attempts to work out the logic of institutions created by the capitalist structure. We will pay particular attention to the status of the labor theory of value within Marx's system, since much of the controversy surrounding Marx as a political economist concerns the adequacy of this aspect of his account.

Economic Reasoning in *Capital*

Let us begin by asking what sort of economic work Marx does in *Capital*. What are his central questions? The following list includes representative research problems associated with his study of the capitalist economy. (It should be noted that these questions pertain to the *capitalist* economic structure; a central part of Marx's view of the historicity of social science turns on his belief that questions of these sorts have different answers in different economic structures.)

1. What is a commodity?
2. How are exchange value, use value, and labor value related?
3. What is the source of profit, interest, and rent?
4. What factors determine the wage rate?
5. What factors determine rates of profit?
6. What factors determine interest and rent?
7. What factors determine the relative prices of commodities?
8. What causes chronic unemployment?
9. What leads capitalist industry to technological innovation and increases in productivity?
10. What factors impose the imperative toward accumulation?
11. What causes the observed tendencies toward concentration and centralization in capitalist industry?
12. What factors influence the long-term behavior of the rate of profit?
13. What are the equilibrium conditions for an economy of several sectors?
14. What factors lead to the business cycle?
15. What factors lead to various forms of commercial crisis?

These are some of the problems Marx raises as a political economist in the classical tradition. They may be grouped under several broad areas: issues in the labor theory of value, problems of profit and price, problems of reproduction, and problems of accumulation. What sort of analysis does Marx provide as a basis for answering each of these families of problems?

The Labor Theory of Value

We may begin with a consideration of the labor theory of value (LTV). The LTV provides Marx's economic idiom, and it is generally taken as a central part of his theory of capitalism. What is the theory of value? What is it supposed to explain? How does it relate to the analysis contained in *Capital* as a whole?

The labor theory of value has these components:

1. A theory of the origin of the value of a commodity (abstract labor time) based on an abstract description of the system of production.
2. A theory of the origin of profits: surplus labor time over and above the time necessary to reproduce the worker's means of life.
3. An incomplete theory of price: Prices correspond to values (with corrections for the conditions of competition).
4. An incomplete theory of the wage: The wage is equal to the value of the means of subsistence, adjusted by class struggle.
5. A derivation of the implications of the theory of value: crisis, concentration and centralization of industry, falling rate of profit, and so on.

Marx's basic hypothesis is that capitalism is a system of production defined by a particular system of property and class. (This is the substance of the description of capitalism in chapter 1.) The LTV is intended to embody these assumptions in a way that permits rigorous economic analysis. Thus *Capital* is the detailed exposition of the dynamic for the economic structure as a whole created by the fundamental relations of production of capitalism. Much of the argument in *Capital* is intended to draw out the implications of the basic structure of the mode of production (through the medium of the theory of value) for the structure, organization, and developmental tendencies of the society in which that mode exists.

Let us begin with a brief sketch of the labor theory of value and surplus value.[1] The most important feature of the LTV is that it is rooted in the technical conditions of production rather than the conditions of exchange. That is, Marx's basic perspective on capitalism is from the point of view of the process of production rather than from that of the competitive markets, financial institutions, and so forth that surround capitalist production. Marx puts this point in the *Grundrisse* (in somewhat Hegelian fashion) in discussing the relations between production, distribution, exchange, and consumption within capitalism. "The conclusion we reach is not that production, distribution, exchange and consumption are identical, but that they all form the members of a totality, distinctions within a unity. Production predominates not only over itself, in the antithetical definition of production, but over the other moments as well" (*Grundrisse*, p. 99). Thus production is primary over consumption, distribution, and exchange. This standpoint reflects Marx's assumption that basic economic parameters are determined by the technical conditions of production: the amounts of labor time, machinery, and raw materials expended and the technique of production employed. Therefore, if we are to understand capitalism as a system, it is first necessary to have a thorough understanding of its process of production.

The labor theory of value maintains that the value of a commodity is equal to the amount of socially necessary labor time expended in producing it. "As crystals of this social substance [homogeneous human labor], which is common to them all, they are values—commodity values" (*Capital I*, p. 128). It is critical to this account that the labor time involved is socially necessary labor time: "Socially necessary labour-time is the labour-time required to produce any use-value under the conditions of production normal for a given society and with the average degree of skill and intensity of labour prevalent in that society" (*Capital I*, p. 129). This requirement blocks the implication that less efficient workers— those who expend more labor time than the industry average per unit of commodity—produce commodities of higher value than their competitors; rather, on Marx's account, some of the inefficient producer's time is wasted and does not contribute to the value of the commodity.

In order to compute the value of the commodity, then, it is necessary to consider the process of production typical of a given industry at a given time and take account of the various sources of labor time expended in the process: the living labor expended by workers and labor expended in the past in the production of the machines, factory buildings, and raw materials. The value of the good is therefore equal to the sum of the sources of labor time expended in its production:

value of commodity $= c + nL$,

where c is equal to the constant capital expended (raw materials and depreciated tools, machines, and buildings), L is equal to the length of the workday, and n is the number of workers).

Let us turn now to Marx's theory of surplus value. All societies produce a surplus product in some form, that is, a stock of goods over and above what is necessary to supply the subsistence needs of the population. The physical surplus takes a variety of social forms, however, in different societies — stocks of wealth held by the Roman slaveholders, rent and labor services collected by the feudal lords, and profit, interest, and rent collected by capitalists. A critical feature of these differences is the means by which a privileged class extracts the surplus from the class of immediate producers.

What mechanism of surplus extraction distinguishes capitalism from feudalism or slavery? This is a central question for Marx, for precapitalist systems of production depended on extraeconomic systems of surplus extraction: the coercive force needed to discipline and control a group of slaves or a hamlet of feudal peasants. Workers, however, unlike serfs or slaves, own their labor power and are free to expend it as they choose. Marx's solution to this problem is the discovery that capitalist surplus extraction works through economic rather than extraeconomic coercion. Capitalism is a system of wage labor: Workers are compelled by subsistence needs to sell their labor power to capitalists at a rate which permits the latter to expropriate the surplus. "For the transformation of money into capital, therefore, the owner of money must find the free worker available on the commodity-market; and this worker must be free in the double sense that as a free individual he can dispose of his labour-power as his own commodity, and that, on the other hand, he has no other commodity for sale, i.e. he is rid of them, he is free of all the objects needed for the realization of his labour-power" (*Capital I*, pp. 272–73). Workers have been separated from the instruments of production through which alone it is possible to transform abstract labor power into useful goods. Consequently they have no alternative but to sell their labor power to the possessor of wealth in order to satisfy subsistence needs. This dual characteristic constitutes the basis of capitalist surplus extraction: Worker are free to sell their labor power to capitalists (i.e., they are not subject to extraeconomic prohibitions against the sale of labor power), and they are compelled to do so because they have no other way of satisfying subsistence needs within the system of capitalist property relations.

Thus the capitalist form of the surplus product – surplus value – originates in the following fact. All the commodities expended in the process of production are assumed to be purchased by the capitalist at their full value. The commodity workers sell is their labor power – their capacity to engage in productive activity for a specific length of time. Labor power is purchased and sold as a commodity within capitalism, and its value is established in the same way as all commodities: by the amount of socially necessary labor time required for its production. In particular, the value of a day's labor power is equal to the value of the wage goods workers need in order to work each day. (This also includes the value of the resources expended on training workers.) The expenditure of labor power within production creates value, and if the length of the workday is longer than that quantity of labor necessary for the production of one day's worth of labor power, then there will be a net surplus. The difference between the length of the workday and the value of labor power is the surplus value created in the process of production:

$$s = L - v,$$

where s is the surplus value, L is the length of the workday, and v is the value of a day's labor power (variable capital).

This account depends on a schematic theory of the wage workers receive; in practice, the wage is more complex than this suggests. In particular, Marx maintains that the wage is typically higher than the biological subsistence level and reflects both biological and "socially determined" needs. "[The worker's] natural needs, such as food, clothing, fuel and housing vary according to the climatic and other physical peculiarities of his country. On the other hand, the number and extent of his so-called necessary requirements . . . are themselves products of history. . . . In contrast, therefore, with the case of other commodities, the determination of the value of labour-power contains a historical and moral element" (*Capital I*, p. 275). The wage is therefore partially determined by class struggle: In periods in which the working class is in a strong bargaining position, the wage will be higher. In general, however, the wage will be kept relatively low because of the inequalities of bargaining strength between workers and employers.[2]

One important function of the labor theory of value is to provide an aggregator in terms of which to describe diverse processes of production, for example, the production of steam engines, cereals, and petrochemicals. The physical tools and machinery, raw materials, and concrete labor skills employed are widely different; consequently it would be convenient to have an abstract quantity in terms of which to aggregate these different factors. The labor theory of value provides such an aggregator; for the process of production in any industry can be analyzed according to the following formula:

$$c + v + s,$$

where c is equal to the value of the constant capital expended in the process of production; v is equal to the value of the labor power expended in the process of production (the value equivalent of the wage); and s is equal to the surplus value appropriated (the difference between the total amount of labor time expended and the value of the labor power expended).

Several ratios are important to Marx's labor value analysis of capitalism. One he refers to as the rate of surplus value or the rate of exploitation (*Capital I*, pp. 320–29). It is the ratio of the surplus labor time to the necessary labor time within the typical workday:

$$s' = s/v.$$

This ratio is important because it is a measure of the efficiency of a given industry as a producer of surplus value and profits. A high rate of surplus value leads to a greater surplus value being created in a given period. The rate of surplus value is determined by two factors: the length of the workday and the value of the wage. Marx refers to this as the rate of exploitation because it gives a measure of the degree to which workers are exploited. It is the ratio of the surplus value expropriated from workers to the part of the workday that goes into satisfying their needs. A high rate of surplus value represents the circumstance in which workers reproduce their wages in a small portion of the workday, while the capitalist expropriates the remainder of the workday; consequently it represents a circumstance in which the workers are more highly exploited.

Another important ratio in Marx's value analysis of capitalist production is the "organic composition of capital" (*Capital III*, pp. 244–45). This quantity is the capital-labor ratio:

$$m = c/v \ (Capital \ III, \ p. \ 245).$$

This ratio characterizes a given industry and represents the proportion of constant capital (machinery, raw materials, buildings) to variable capital (wages). (Note that this ratio is distinct from the "technical composition of capital," which is the proportions of concrete goods and labor power involved in the process of production; *Capital III*, p. 244.) One of Marx's central arguments in *Capital* is that the organic composition of capital tends to rise as a result of the imperative toward technological innovation. (See chapter 5 for extensive discussion of this argument.)

These comments suffice to establish the basic features of the labor theory of value. It should be remarked, however, that up to this point we have seen no substantive study of particular economic problems. Rather, we have examined a framework of description and analysis: an abstract system of description of the process of production in terms of embodied labor time. And we have identified two hypotheses that motivate the choice of this system of analysis: Marx's assumption that the technical features of production are economically decisive and his hypothesis that the specific system of surplus extraction characteristic of a society is explanatorily fundamental.

Profits and Reproduction

Let us now turn to some of the substantive problems to which Marx directs his economic research. First, consider the central issue of profit. Marx's theory of profit is founded on his theory of surplus value, which maintains that profits originate in the surplus labor performed by wage laborers. On this account the total sum of profits is equal to the total surplus labor time appropriated during production. Thus Marx offers an analysis of profit based on the technical conditions of production, not primarily on conditions of exchange (market relations).

The rate of profit may be represented as follows:

$p' = s/(c + v)$ (*Capital III*, p. 133).

This expression may be transformed by dividing numerator and denominator by v; we then arrive at the following expression:

$p' = s'/(m + 1)$,

where s' equals the rate of surplus value (s/v) and m is the organic composition of capital (c/v). Crucial to determining the mass and rate of profit, then, within a capitalist industry is the rate of surplus value: the proportion of surplus labor time to necessary labor time within the production process (s/v). The mass of profit equals the total surplus value expropriated, and the rate of profit is directly proportional to the rate of surplus value.

These points summarize the essence of Marx's analysis of profit in terms of surplus value. However, this analysis is wholly formal up to this point; it has only the most meager consequences for actual capitalist phenomena. This formality derives from the fact that profits—both the rate and mass of profit—are observable quantities within capitalism; and they are influenced by factors that are deliberately disregarded within the pure theory of value: factors imposed by the requirements of competition. "The rate of profit is the historical starting-point. Surplus-value and the rate of surplus-value are, relative to this, the invisible essence to be investigated, whereas the rate of profit and hence the form of surplus-value as profit are visible surface phenomena" (*Capital III*, p. 134). To apply this analysis to real problems of the capitalist economy, therefore, it is necessary to consider these interfering factors.

Marx turns to a more concrete treatment of profit in volume III of *Capital*.

Let us consider an important illustration from this treatment: Marx's effort to take into account the effects of competition on the rate of profit. Here the problem is the fact that according to the theory of value, industries with different organic compositions of capital should possess different rates of profit. Consider two different industries that employ the same number of workers (variable capital) but possess different compositions of capital. This assumes that one industry is more capital intensive than the other. If the rate of surplus value is constant in both industries, both will produce the same quantity of surplus value. The capital-intensive industry puts in motion a larger amount of total capital; its rate of profit thus must be lower than that of the labor-intensive industry. But this outcome is impossible in a competitive economy, since capital would migrate from the low-profit to the high-profit sector (*Capital III*, pp. 245–49). Consequently Marx is compelled to provide an account of the mechanisms by which an average rate of profit is created for the economy as a whole. "The object of this Part is simply to present the way in which a general rate of profit is arrived at within one particular country" (*Capital III*, p. 242).

For industries with different organic compositions to possess the same average rate of profit, it is necessary to alter the assumption that commodities are sold at their value. Consequently Marx introduces the concept of the price of production, that is, the price at which the commodity must be sold in order to generate a return on total capital invested that equals the average rate of profit. "The prices that arise when the average of the different rates of profit is drawn from the different spheres of production, and this average is added to the cost prices of these different spheres of production, are the *prices of production*" (*Capital III*, p. 257). On this analysis, individual capitalists no longer expropriate all and only the surplus value created within their own units of production; rather, the surplus is effectively aggregated through the competitive system for the capitalist class as a whole and then returned to individuals in proportion to the size of the capital investment each makes (*Capital III*, p. 258).

After introducing this qualification to the fundamental theory of profit as surplus value, Marx returns to the theory of value and asks how the rate of surplus value affects the rate of profit under these new circumstances. Here he finds that the law of value remains relevant in several ways. First, within a given industry a change in the amount of labor time needed in the process of production will cause the price of production to change and consequently will affect the rate of profit. Thus the technical features of production, which the labor theory of value is intended to express, continue to determine the rate of profit, though in a more complicated fashion. Second, Marx asserts that this account shows that aggregate profits equal aggregate surplus value; all that has changed is that individual profits no longer equal individual surplus value (*Capital III*, p. 280). (It is now known that the latter assumption is not quite correct: To solve what is now called the transformation problem, it is necessary to give up the assumption that net profits equal net surplus value. More radically, Ian Steedman argues that the problem is both insoluble and unnecessary.)[3]

This body of analysis represents Marx's effort to cope with a range of complex phenomena within a capitalist economy: the effect of competition on the rate of profit in different sectors of the economy. Other instances of Marx's work in this area could be considered as well, for example, his account of the influence of the period of circulation of commodities on the rate of profit (*Capital II*, pp. 207–14), his analysis of the effects of changes factor costs on the rate of profit (*Capital III*, pp. 200–19), and his discussion of the division of the social surplus among different segments of the capitalist class (*Capital III*, parts 5 and 6). Critics differ on the degree of Marx's success in this area, but it is clear that he is engaged in substantive economic analysis in these chapters and not merely providing a formal elaboration of the labor theory of value. This work therefore represents a significant addition to our understanding of an important aspect of capitalist production.

Let us now consider a second large body of economic analysis in *Capital*: Marx's use of "schemes of reproduction" to represent the flow of commodities through the capitalist economy. The fundamental economic problem here is that for commodities to be sold there must be sufficient effective demand for them, and as demand for certain kinds of commodities falls, demand for other types of commodities will fall as well. For example, if demand for woven cotton goods drops, this will lead ultimately to a decline in the effective demand for looms as well. The latter decrease will lead to lowered wages as loom manufacturers lay off workers, which will lead to a further decrease in demand for cotton goods. It is thus possible for crises of disproportion to arise in which some industries have overproduced relative to demand, leading to a spreading series of unsold inventories. In volume II Marx attempts to provide a two-sector model of the economy as a flow of goods and labor time among industries that will establish a set of equilibrium conditions of a capitalist economy. As a simplifying assumption he stipulates that profits are spent in luxury consumption rather than productive reinvestment. This condition of equilibrium is what Marx refers to as simple reproduction; in later chapters he describes the conditions necessary for the smooth expansion of capital investment, which he calls expanded reproduction.

According to Marx's model we can aggregate all industries into one of two sectors: capital-goods producers and consumption-goods producers. "The society's total product, and thus its total production process, breaks down into two great departments: I. Means of production. . . . II. Means of consumption" (*Capital II*, p. 471). Some industries produce commodities used chiefly for consumption, such as clothing and food; others produce commodities used chiefly for production, for example, steam engines and rolled steel. This distinction concerns the nature of the output of the productive process, but we also need information about the forms demand takes within capitalist production. Effective demand requires a source of income, and at the highest level of abstraction there are only three sources of income within the capitalist economy: income accruing to workers through wages, to capitalists through profits, and to capitalists in the form of replacement costs of depreciated capital. This analysis of income corresponds exactly to the value description of the process of production: $c + v + s$. Thus at the end of a period of production, capitalists have money in the amount of $c + s$, and workers have money in the amount of v (*Capital II*, pp. 471–72).

Finally, we can inquire where these funds are to be expended. We find that capitalists will expend c in Department I (capital goods), and workers will expend v in Department II (consumption goods). On the assumption of simple reproduction (i.e., no productive reinvestment of profit), capitalists will expend all of s in Department II (consumption goods). (When we get to the stage of expanded reproduction, s must be divided between Departments I and II.)

These suppositions establish the conditions of equilibrium for a capitalist economy on the assumption of simple reproduction: The output of Department I must be equal in value to the total constant capital consumed in a given period in the economy as a whole, and the output of Department II must be equal in value to the total sum of variable capital and surplus value generated in the economy as a whole. Finally, all industries have been aggregated into one department or the other. Therefore the value of Department I must equal the capital expended in Departments I and II, and the value of the output of Department II must equal the sum of $v + s$ for both departments. Thus we arrive at these equalities (*Capital II*, pp. 474–87):

$$c_I + v_I + s_I = c_I + c_{II}$$
$$c_{II} + v_{II} + s_{II} = v_I + v_{II} + s_I + s_{II}.$$

These equalities imply the following condition of equilibrium:

$$c_{II} = v_I + s_I \ (\textit{Capital II}, \text{ p. } 478).$$

That is, the total capital consumed in the consumption-goods sector must equal the total consumption demand created in the capital-goods industry; in other words, exchange between the two sectors must be equal. If there is an excess on either side, then a crisis of circulation will develop: Goods will be produced that cannot be sold.

This body of analysis provides a device for describing the equilibrium conditions the system must satisfy for simple reproduction of capitalist production. Marx's schemes of reproduction are useful for several reasons. First, they offer a way of characterizing the movements of value and goods within the economy as a whole—a form of analysis that puts Marx squarely within the tradition extending from the Physiocrats to Leontief's input-output analysis. Second, they demonstrate an important source of crisis within capitalism: crises of disproportion between departments, leading to a spiral of inventories that cannot be sold. (It should be remarked that this model uses the concept of value as embodied labor time as an aggregator, but it is otherwise independent of the LTV.) Michio Morishima has high regard for this construction.[4] It offers an explanation of many features of the capitalist economy, including some aspects of crises. Other economists (e.g., Ian Steedman) argue that labor time is not an adequate aggregator and that physical quantities are superior.[5] Whatever we conclude here, the model itself is an important technique of analysis.

There are several connections between the economic work described here and the institutional account described in chapter 1. First, Marx believes that the capitalist economy works as it does *because* of the social institutions that define it. The details of the economic analysis rest on substantive institutional assumptions. In particular, private property and class coercion are the prerequisites of production of commodities for profit. Reversing this point, if the social institutions of private property and class were undermined, the economy would falter. (For example, large numbers of workers would withdraw from the pool of wage labor, driving wages higher than profits could absorb.) Second, Marx believes that the LTV expresses these institutional prerequisites of capitalism in an especially natural way. The theory of value itself represents the central place of the immediate producer within a system of wealth production. And the theory of surplus value represents the coercive structure of the system of private property: The bourgeoisie compels the proletariat to perform unpaid labor. Thus the labor theory of value expresses the class nature of the system.

It should be noted, however, that much of Marx's substantive economic analysis is independent of the labor theory of value; it is possible to recast his basic arguments in terms that do not depend on labor value. The LTV usually plays an important role in Marx's account, but its importance varies from case to case. In fact, it will be argued later that the LTV can be eliminated from most of Marx's economic arguments without vitiating them.

Competition and Price

Let us turn now to a second level of Marx's analysis of capitalist production, the level of price and profit. These are the terms in which the capitalists' own calculations are performed. Each of the central concepts of the theory of value has a corresponding category at the level of competition: value and price, surplus value and profit, rate of surplus value and rate of profit, organic competition of capital and capital-labor ratio computed in price terms, and so on. The money-price description of capitalist production constitutes the observational level of the phenomena singled out in the theory of value, and it reflects the conditions imposed by the fact that capitalism is a system of competition within a market. Capitalists' accounting is done in money terms; they are concerned with the prices of raw materials and labor, and with the ultimate price of the commodity they can obtain at the time of sale. The price of the commodity is determined in the short run by the balance between supply and demand. Profit is calculated in terms of the difference between the net money cost of the commodities necessary for the production of the new product (rent, interest, raw materials, machinery, and wages) and the net money price that can be realized for the finished product.

Marx relates these two levels in several ways. First, he argues that the description of the economy in price terms cannot be ultimate.[6] The laws of competition determine the fluctuations of prices from day to day, but they do not explain the equilibrium points around which prices fluctuate. In order to explain those points, it is necessary to study the process of production that permits us to account for the real determinants of price in terms of rational investment and disinvestment. The theory of value gives us such an analysis. Having described the process of production in value terms, Marx is able to show that declines in the market price of the output cause the price of the good to fall below its true value, thereby reducing the amount of surplus value and in turn reducing the rate of profit. The fall in the rate of profit gives capitalists an incentive to shift their investment to an industry in which the rate of profit is higher. This reduces the supply of goods in the original industry, which in turn tends to bring the price of the commodity back to its real value. On this first attempt to relate prices to values, then, Marx suggests that values are the points around which prices fluctuate, and the theory of value explains why these are the price equilibria.

This account does not quite suffice, however, since the laws of competition intervene in another way as well. The theory of value establishes that different industries commonly have different organic compositions of capital: different ratios of industrial capital to the wage bill. Assuming a fixed rate of surplus value, the organic composition of different industries implies a difference in the rates of profit in different industries as well. (The rate of surplus value is the ratio of surplus value to the wage bill, whereas profit is the ratio of surplus value to the sum of the wage bill *and* the industrial capital.) The basic conditions of competition, however, require a single uniform rate of profit common to all industries; otherwise, investments would leave low-profit industries and flow to high-profit industries. This movement, however, would increase the supply of the high-profit commodity and therefore reduce its price; this in turn would reduce the rate of profit. Competition ensures, then, a uniform rate of profit across industries.

Marx regards the conditions of competition to be of subordinate theoretical significance (although certainly a defining feature of a capitalist economy). For this reason, he puts off the analysis of competition until the later volumes of *Capital*. Having taken account of the conditions of competition, however, we can see that the relation between values and price is more complicated than the first treatment allowed. (It is worth noting that this complication was well known to Marx when he wrote volume I of *Capital*: in fact, volumes II and III of *Capital* were already in manuscript form before volume I was written.) This complication (the transformation problem)[7] can be accommodated by the introduction of a new concept: the cost of production. It is possible to calculate the ideal prices of commodities in industries of different organic compositions, given their true values and the condition that the rate of profit must be uniform. Having provided this construction of the concept of the cost of production, Marx's view of the relation between prices and values stands complete: Values plus the conditions of competition establish costs of production, and prices fluctuate around these costs.[8]

It should be noted that Marx is not primarily interested in providing a theory of the determinants of price or in working out the dynamics of the competitive market. Rather, his aim is to provide an analysis of the capitalist economic structure that will explain the large-scale dynamics of that system. And the labor theory of value is used to express the economic relations that give rise to the dynamic properties of capitalism. Allen Wood makes this point when he writes, "The law of value is a proposition of economic science, employed to explain what actually happens in capitalism. But . . . it is not intended as a theory of relative prices for capitalism, but only as a postulate for the basic and extremely abstract model of commodity exchange with which Marx's dialectical theory begins."[9] This is not to say that Marx's account does not have implications for the theory of price, but only that its intended application is to the macrophenomena of capitalism rather than the microphenomena.

The Role of the Labor Theory of Value in *Capital*

Value and Explanation

The labor theory of value serves several different functions for Marx, the most important being that it provides a way of analyzing the process of production independently from the phenomena of competition and price. Any system of production can be viewed abstractly as a set of institutions which take various forms of social wealth as input and provide other forms of social wealth as output. These include raw materials, tools, factories, land, knowledge, and labor power and skills. (Piero Sraffa's phrase "the production of commodities by means of commodities" puts the point nicely.)[10] When goods are purchased and sold within a market economy, the buyer and seller perceive only the money price of the good; but one of Marx's central insights is that the money price must reflect the technical characteristics of the production process. That is, a commodity produced in a process that consumes large amounts of social wealth per unit ought to have a high price, and vice versa. The problem confronting economists is to provide an abstract way of characterizing the process of production that will illuminatingly reflect the flow of social wealth through the process.

The labor theory of value is just such an effort. It is a system of measurement for describing the movement of social wealth and commodities throughout the capitalist economy. By singling out abstract labor time it is possible to describe the process of production as an organized flow of labor time. Different industries produce qualitatively different goods (as a consequence of different forms of concrete labor and concrete capital), but in value terms, they all take certain quantities of capital (frozen abstract labor), mix them with quantities of living abstract labor (wage labor), and produce commodities having a definite quantity of abstract labor congealed within them. The labor theory of value, therefore, affords a system of description in terms of which Marx can describe the organization of the economy as a whole. Value serves as an aggregator, a way of comparing the organization of different firms and different industries – qualitatively incommensurable industries.[11]

Having provided a value description of the capitalist mode of production (which in principle can be done for *every* system) Marx makes a further supposition; namely, that the principle of value *regulates* the process of capitalist production. According to the theory of value, surplus labor time is the sole source of profit. Capitalists make decisions on the basis of prices, profits, rates of interest, rents, and so forth. However, firms that fail to organize their production in such a way that surplus value is produced in sufficient quantities will not show a profit and will be forced to withdraw from production. This does not mean that capitalists conduct their calculations at the level of value rather than price. It merely means that however they organize production and calculate the appropriate use of resources, their methods must implicitly conform to the requirements of value and surplus value or the enterprise will not be profitable. In this way, then, capitalism is regulated by the principle of value: The theory of value establishes a set of conditions the successful capitalist enterprise must satisfy in one way or another.

Finally, Marx argues that the value description of the capitalist mode of production has institutional consequences for the development of the economy as a whole, and he devotes considerable effort to uncovering those consequences. These are the "laws of motion" Marx set out discover. They include the law of the tendency of the rate of profit to fall; the formation of an industrial reserve army; the tendency to concentrate and centralize within capitalist production and ownership; and the tendency toward various forms of economic crisis. These implications will be discussed in greater detail in chapter 5.

Thus the theory of value is explanatory because it describes the capitalist economy from the point of view of the process of production. It identifies the underlying functional requirements of the system of production around which the economy is organized. Marx's basic insight is valid whether or not we accept the particulars of the labor theory of value: the technical features of the process of production determine prices and profits. If innovations result in a dramatic drop in the quantity of social resources needed to produce a good, then ultimately the price of the good must drop, given conditions of competition. The theory of value contributes to an explanation of the process of production because it identifies the true social costs of production. It analyzes the economy in terms of a flow of social wealth—raw materials, labor, land—in abstract terms: embodied labor time.

Recent Criticisms of the Labor Theory of Value

Many critics have argued that the LTV is a handicap to the scientific standing of Marx's theory of capitalism: that it is not a useful tool of economic analysis,[12] that it relies on an occult notion of value,[13] and so forth.[14] Joan Robinson expresses a common view when she writes that "no point of substance in Marx's argument depends on the labour theory of value."[15] Two sorts of objections have been raised against the LTV. The first concentrates on a long list of technical problems internal to the logical development of the theory. Criticisms of this sort go back at least as far as Böhm-Bawerk, and they include the transformation problem, Marx's assumptions about the rate of surplus value, and his assumptions about change in the organic composition of capital. A second family of objections centers around the claim that the LTV is unnecessary altogether. Marginalist economists have long contended that the labor theory of value has been replaced by the utility theory of value developed by Marshall and others. And more recently, Ian Steedman has applied Piero Sraffa's critique of marginalism to Marx, arguing that Sraffa's system demonstrates the redundancy and inadequacy of the LTV.[16]

In this section I will offer some support for many of these criticisms, without drawing the destructive conclusions such arguments often suggest. I will maintain that Marx overestimated both the explanatory power and the theoretical importance of the labor theory of value and that a balanced appraisal of his system will attach greater importance to the substantive institutional assumptions described in chapter 1, and the concrete economic analysis given here, and less to the LTV. However, I will argue that the LTV does serve an important function in Marx's analysis: It directs attention to the technical organization of the process of production rather than to the conditions of circulation. Marx brilliantly establishes the importance of this dimension of a theory of capitalism; and the question of whether to examine the process of production in value terms or in physical parameters is less important. Moreover, I will argue that almost all of Marx's arguments survive this revaluation of the status of the theory of value.

Michio Morishima's *Marx's Economics* is an especially important treatment of Marx's economic account. Morishima provides an extended examination of Marx's main economic doctrines using modern mathematical techniques. He constructs a rigorous statement of the labor theory of value and uses this to investigate Marx's view of exploitation, the rate of profit, the transformation problem, reproduction, and other central problems in his economic system. Morishima ultimately concludes, however, that the labor theory of value in its classical statement is irredeemable and must be replaced.

Morishima gives Marx a high rating as a mathematical economist. "Marx . . . should in my opinion be ranked as high as Walras in the history of mathematical economics. . . . Indeed, Marx's theory of reproduction and Walras' theory of capital accumulation should be honoured together as the parents of the modern, dynamic theory of general economic equilibrium."[17] One of Marx's most important achievements, according to Morishima, is his dynamic model of the economy (his reproduction schemata) and the consequences which he draws from this model. Moreover, the LTV plays an important part in Marx's presentation of this model:

> The labour theory of value plays a most important part in Marx's economics, since it provides a system of constants, in terms of which his microeconomic model may be aggregated into a two-departmental macroeconomic model.[18]

Among Marx's economic achievements Morishima includes his rigorous macrodynamic model (the schemes of reproduction, which Morishima sees as anticipating Leontief's input-output analysis), his derivation of some dynamic laws of the CMP, his account of the capital-labor ratio (the organic composition of capital), his idea of the duality between value and price systems—"these would be enough examples to recommend Marx as a purely academic economist for one of the very few chairs with the highest authority."[19]

Much of Morishima's book is a mathematical development of some of Marx's central ideas in the theory of value. He provides a sophisticated technique for representing labor-time contents (values) in a commodity-producing economy and uses this technique to derive several important theorems. Especially important is his proof of a proposition fundamental to the theory of exploitation. Morishima shows that the rate of profit is positive if and only if the rate of exploitation is positive.[20]

At the same time, Morishima concludes that Marxian economics must be pruned if it is to remain a serious effort in mathematical economic analysis. Morishima believes that the breakdown theory (the idea that capitalism must necessarily collapse under accumulating crises) must be qualified;[21] but more important, the labor theory of value must be dramatically altered under pressure from several technical considerations: the heterogeneity of labor, problems arising from joint production, and problems arising from alternative techniques of production. These factors entail that there will be no unique positive set of values to assign to the commodities produced in a given period of production. In Morishima's account, however, where the chief function of the LTV is to serve as an aggregator, this failing is fatal.

Thus Morishima's position comes down to this: The LTV was an insightful technique of aggregation that permitted Marx to derive important conclusions about the behavior of the CMP. It was a useful first approximation to economic processes that can now be investigated more fully with more sophisticated schemes of aggregation.[22] The LTV is now known to have serious technical defects and ought to be replaced by more sophisticated techniques of aggregation. Finally, however, Morishima believes that the main economic doctrines of Marx's theory survive this surgery.

Another recent controversy over the labor theory of value is the critique of the LTV based on the work of Piero Sraffa. Sraffa's *The Production of Commodities by Means of Commodities* is an analysis of capitalism in the tradition of Ricardo, Quesnay, and other classical political economists. The account is intended to serve as the basis for a systematic criticism of marginalist theory. Sraffa's technique consists in depicting the economy as a set of industries represented as linear input-output functions that take certain physical quantities as parameters (raw materials, tools, labor time, etc.). Each industry absorbs a given mix of physical commodities and labor power (fixed by existing techniques of production) as inputs and provides a given quantity of new commodities as outputs. Using this information and information about the real wage going to laborers (the quantities and types of physical goods included in the wage basket) Sraffa shows that it is possible to generate a determinate set of prices and profits. Therefore no theory of value is needed; prices and profits can be explained using only physical quantities of goods and labor power.

Ian Steedman's *Marx After Sraffa*[23] extends this analysis to the problems Marx treats with the LTV and shows (1) that Sraffa's technique permits derivation of both value quantities and price-profit data, thus proving that the physical-production data is more fundamental than labor-time content; and (2) that value data does not permit the derivation of price-profit data.[24] This latter point derives from Steedman's discussion of Marx's treatment of the transformation problem.

Thus Steedman concludes that the Sraffa technique demonstrates the insufficiency and redundancy of the LTV. He does not regard this as a crippling conclusion for Marxist economic accounts, however, because he distinguishes between two parts of classical political economy that are usually conflated by neoclassical critiques of Marx. Classical political economy is a "surplus-appropriation" account of capitalism in that it centers its analysis on the production of a surplus quantity of goods in the process of production over and above the replacement of capital goods and wage goods.[25] Second, classical political economists couched their analysis largely in terms of the labor theory of value. Marx, along with other classical political economists, believes that the LTV substantiates the surplus-appropriation approach in that the LTV demonstrates the origin of the surplus (surplus labor time). Neoclassical political economy, on the other hand, rejects the surplus appropriation model implicitly through its rejection of the LTV. Steedman argues, however, that Sraffa's technique represents a third position, one based on examining the economy as a system of surplus appropriation while at the same time rejecting the LTV.[26] Thus Sraffa demonstrates the independence of surplus-appropriation accounts from the LTV. Steedman maintains that the former constitutes the real core of Marx's study since it permits analysis in terms of class and exploitation. Steedman therefore regards his critique as a first step toward a more adequate Marxist surplus appropriation account that is not clouded by reliance on the LTV.[27]

Further Grounds for Doubting the Importance of the LTV

Morishima's and Steedman's arguments are internal to the LTV in the sense that each contends on technical grounds that the LTV is defective for use in economic analysis. But the same conclusion can be reached from a different direction. I will argue that the LTV is generally dispensable in Marx's *own* explanations—even though his own formulations are usually given in value terms. Most of the "central tendencies" Marx attributes to the capitalist mode of production are based on reasoning about rational behavior within the institutions that define the capitalist mode of production, without making essential use of the concept of value. Consequently these arguments, which constitute much of the explanatory power of Marx's account of capitalism, are independent of the LTV. This finding gives us strong reason to suppose that the labor theory of value is in principle dispensable. And it shows that Morishima's and Steedman's criticisms can be accepted without serious harm to the main elements of Marx's economics, since Marx's own account of capitalism can be put in nonvalue terms. To establish this independence, we must consider some of Marx's central arguments and show that they may be formulated in nonvalue terms.

Consider first Marx's discussion of the imperative toward accumulation within capitalism. This discussion is contained in chapter 25 of *Capital*, "The General Law of Capitalist Accumulation." Marx posits that capitalists are subject to a powerful compulsion to accumulate capital, that is, to invest a large portion of profits in new productive capital. In chapter 25 Marx tries to uncover some of the social and economic consequences of this imperative. Let us begin, therefore, by asking why he believes this assumption to be true. This question—why must capitalists accumulate capital?—has two aspects. First, we may ask what compels a given capitalist to accumulate within typical capitalist institutions. The answer is that other capitalists *are* energetically accumulating, which would give them a potentially fatal competitive advantage. Therefore, nonaccumulating capitalists, recognizing the possibility of extinction, must change their ways. This compulsion is reminiscent of Weber's description of the capitalist in an iron cage: Capitalists cannot escape the imperatives created by capitalist institutions.

> The Puritan wanted to work in a calling; we are forced to do so. For when asceticism was carried out of monastic cells into everyday life, and began to dominate worldly morality, it did its part in building the tremendous cosmos of the modern economic order. This order is now bound to the technical and economic conditions of machine production which today determine the lives of all the individuals who are born into this mechanism. . . . Perhaps it will so determine them until the last ton of fossilized coal is burnt.[28]

Note that this reasoning depends only on the conditions of choice imposed on individual capitalists by the structure of capitalist production. The labor theory of value has no role here at all.

Second, we might ask what creates the impulse to accumulation in the first place. Here Marx's account is less clear. Accumulation is one of the preconditions of capitalism, not a consequence of some more basic factor. In particular, production for profit is compatible with nonaccumulation, if every capitalist is content with his own share. And this assumption may be satisfied in societies regulated by traditional values (e.g., fifteenth-century Italy). (This is one of the central problems of Weber's *Protestant Ethic*: What sociocultural change in individual motivation can account for the creation of an economic system based on accumulation in Western Europe?) However, such a system becomes unstable if we add the assumption that every capitalist energetically seeks to *maximize* the rate and amount of profits under competitive conditions. For each capitalist can increase the total amount of profits by increasing the scale of production, which requires increased investment in each cycle. Once this inducement is generally recognized, it becomes mandatory for every capitalist (even those who are satisfied with the current level of profits) for the reasons given earlier: They cannot be certain that their current level of production will allow them to continue to compete. But this account does not establish that owners of wealth will inevitably seek to maximize profits—and if they do not, there will be no resulting impulse toward accumulation. Note, finally, that the reasoning offered here is wholly independent of the theories of value and surplus value.

This independence of the laws of accumulation from the labor theory of value is characteristic of others of Marx's arguments as well. Let us now consider his account of the creation of an industrial reserve army. He first maintains that the imperative toward accumulation leads to a constant striving for more effective techniques of production. This movement leads to a rise in the organic composition of capital since as capitalists seek improved techniques, they tend to apply more advanced and more extensive technology (*Capital I*, pp. 772–73). But as techniques are introduced which permit a smaller labor force to produce more goods, there is a declining demand for wage labor. Relatively fewer workers are needed, leaving others unemployed. Marx represents this process in the form of a change in the value composition of capital: The value of constant capital tends to rise as a fraction of the value composition of the commodity.

This dynamic (toward more productive techniques based on the application of technology) thus leads to periodic revolutions in the structure of the capitalist economy, as new techniques replace old. Workers with labor skills appropriate to the earlier techniques are expelled, and a new body of workers with skills appropriate to the new technique is recruited. And given the diminished demand for labor created by the new technology, the number of workers rehired may be expected to be smaller than the number expelled.

This argument depends chiefly on assumptions about the technical implications of the drive for higher productivity. The application of technology to the production process may be expected to increase the capital-labor ratio. But there is another dimension of the argument: When the capitalist has a choice between two comparably efficient techniques of production, he has a class incentive to choose the capital-intensive alternative, since this puts him in a stronger bargaining position with the labor pool.

For reasons of this sort, Marx argues that the process of accumulation tends to create a pool of "surplus" workers. This pool depresses the level of the wage and consequently increases the rate of profit. Once again, it is apparent that this account is not intrinsically dependent on the LTV. Rather, it relies on Marx's analysis of the institutional and class circumstances within which capitalist managers make decisions, and the implications these conditions of choice have for the economy as a whole. Marx's account is given in value terms, but nothing essential depends on that formulation. We would find a similar degree of independence in looking at Marx's discussion of other implications of the drive for accumulation: concentration and centralization and the tendency toward crises based on the devaluation of capital and existing stocks of finished commodities.

Consider finally Marx's argument about the falling rate of profit. We maintained earlier that Marx seeks to show that the organic composition of capital tends to rise as capitalists search for more productive techniques. The organic composition is a value ratio:

$m = c/v;$

it is the ratio of constant capital to variable capital. On Marx's theory of profit, the rate of profit is equal to:

$p' = s/(c + v),$

where s is the total surplus value, c the total constant capital, and v the total variable capital. If we divide numerator and denominator by v, we arrive at:

$p' = (s/v)/[(c/v + (v/v)]$
$p' = s'/(m + 1),$

where s' is the rate of surplus value (or the rate of exploitation). Marx assumes that s' is constant; he has already state that m tends to fall; and therefore p' (the rate of profit) will tend to fall as well.

This argument is more dependent than others on the LTV. In particular, it assumes that total profits are equal to total surplus value. If we reject the theory of surplus value, then this reasoning has no force, since p' will have no relation to the real rate of profit. However, Steedman shows that the argument can be recast in physical unit terms and that similar (though somewhat weaker) conclusions follow.[29] his analysis shows that the argument relies on Marx's "surplus appropriation" assumptions rather than on the particulars of the LTV.

These examples indicate that Marx's laws of motion of capitalism can be derived from general facts about the capitalist mode of production—private property, class, wage-labor, competition, profits, accumulation—without essential reference to the LTV. This in turn suggests that the LTV is theoretically dispensable for the best of reasons: It can be replaced by other tools of analysis without significantly altering the theoretical conclusions of the analysis.

The Status of the LTV in Marx's System

These arguments suggest that the LTV is less central to Marx's account of capitalism than he believed. It can be eliminated from Marx's own favorite explanations (with a few exceptions). In fact, these arguments show that the labor theory of value is not a *theory* at all, but rather a framework of description. There are alternative descriptive frameworks that would work equally well; the only substantive requirement is that the descriptive system allow us to analyze the process of production adequately and demonstrate the conditions the process imposes on economic variables (price, profit, interest, etc.). This substitutability demonstrates that the LTV functions as an analytic tool rather than as a theoretical premise. It is more akin to a set of coordinates than to an empirical hypothesis.

What is left of Marx's theory if we reduce the theoretical importance of the LTV? Marx's theoretical premises were outlined in chapter 1 above; they are the structural and functional characteristics of the capitalist mode of production. Furthermore, the economic analysis Marx bases on those premises can be reformulated in nonvalue terms. Thus reducing the role of the LTV does not seriously diminish the scientific import or scope of Marx's system. His account in *Capital* takes two forms: a description of the distinguishing structural and functional characteristics of the capitalist system of production and the institutional logic they impose on the system, and a technical economic analysis of the properties of the capitalist system of production and exchange. Marx uses the LTV in order to draw these implications, but alternative descriptive schemes exist that would leave the system intact.

Moreover, even if we conclude that Marx overestimates the importance of the LTV, the central on insight which it rests – the centrality of the process of production within a theory of capitalism – remains valid. Sraffa's system reinforces this conclusion by showing that the phenomena of competition (profits and prices) are determined by the technical characteristics of the production process.

The LTV and Exploitation

If the LTV is not an indispensable part of Marx's theory of capitalism, then how can we account for his deep commitment to it? In part this attachment can be attributed to the research program within which Marx locates himself; the labor theory of value is one of the central tenets of the tradition of classical political economy. But there is an additional motive, which derives from Marx's view of the exploitative nature of capitalism. Marx believes that the theory of surplus value provides scientific content to the idea of exploitation by showing that profits are possible only through the expropriation of the surplus labor time provided by the wage laborer. But dispensing with the LTV does not entail loss of the idea of exploitation. Marx's fundamental assumptions about the institutional features of the capitalist economic structure reveal capitalism as a class society based on coercive relations of power and authority that lead to privilege for some and deprivation for the mass of society. This economic structure is defined by inequalities of wealth and power in which those with little wealth and power must submit to a form of production and life designed and controlled by others. Such a system can properly be said to be exploitative – whether or not we adhere to the labor theory of value. Value and class are separable, and the idea of exploitation survives the reinterpretation of the theory of value proposed here. The heart of the theory of exploitation rests with the nature of the property relations within capitalism and the relations of power and authority through which the fruits of productive activity are distributed and enjoyed – not with the theory of value.

John Roemer's work in the theory of exploitation provides strong support for this position.[30] Roemer notes the underlying motive for preferring the LTV as a system of accounting in its apparent support for the interpretation of capitalism as exploitative, and he shows that the latter conclusion can be made without reference to the LTV. "Why do Marxists choose labor power as the numeraire commodity for defining value and exploitation? . . . There is nothing objectively correct about the labor theory of exploitation, in the sense of its being deducible from economic data. It is rather a particular theory of exploitation that corresponds to the interpretation of capitalism as a class struggle between poor workers and rich capitalists, which according to historical materialism, is the most informative historical interpretation of capitalism."[31] That is, the choice of a labor theory of exploitation is determined not by a body of empirical data but rather by a higher-level theoretical premise: the analysis of capitalism as a class system based on a particular set of property relations.

However, Roemer's work shows that the category of exploitation may be directly linked to the character of the property relations in a given society, without the intervening step of the labor theory of value. He offers a definition of capitalist exploitation: "the appropriation of the labor of one class by another, realized because of their differential ownership of or access to the nonhuman means of production."[32] And, crucially, this conception of exploitation is independent of the labor theory of value. "It is possible . . . to characterize Marxian exploitation using solely concepts of property relations . . . without reference to the concept of surplus labor and therefore to the labor theory of value."[33] In particular, Roemer argues that exploitation occurs in every society in which there is differential ownership of property; it occurs as a result of the differential distribution of the means of production across distinct groups in society. Thus "the relevant coercion, as concerns exploitation, is in maintaining property relations, not in the labor process."[34] That is, capitalism represents a system of property relations in which a minority group owns and controls the forces of production; these property relations are coercively enforced (so as to prohibit spontaneous redistribution), and as a result the producers are compelled to transfer to the owners a portion of the product in order to gain access to the forces of production. The result is a set of economic relations in which there is a net flow of commodities from the producers to the owners. Whether this happens through a labor market or a credit market (in the example Roemer develops as an alternative to the classical Marxian account), and whether it is described in terms of a flow of value, a flow of money, or a flow of physical goods, it remains an exploitative relation between owners and nonowners because those who do not own property receive less than their per capita share of the fruits of productive activity, while those who own more receive greater than their per capita share. "I believe that workers are exploited in the Marxian view not because they contribute surplus labor in production, but, more fundamentally, because they do not have access to their per capita share of the nonhuman, alienable means of production."[35]

Roemer thus provides support for the interpretation of Marx's system advanced in this chapter and in chapter 1. The central elements of Marx's view of capitalism are to be found in his description of the property relations that characterize the capitalist economic structure. These relations establish the coercive framework within which productive activity proceeds under capitalism, and they define the distinctive system of incentives and prohibitions that confine the participants (as workers or owners). These property relations give rise to an economic logic that can be examined using a variety of techniques – the labor theory of value, input-output analysis, or cost of production analysis. And the macro-characteristics of this system of property relations – its tendencies of development, the patterns of distribution it spawns, and the degree to which it is exploitative – are robust with respect to the technique of economic analysis. The bulk of Marx's conclusions can be reached using any of these alternative techniques.

4

Essentialism, Abstraction, and Dialectics

So far we have been primarily concerned with Marx's actual practice as a social scientist; we have inferred various features of his theory of science from the explanations and analyses to be found in *Capital*. It is also true, however, that Marx puts forward a number of explicit methodological views throughout his mature work, and one might suppose that these remarks would constitute the primary texts in which to locate his theory of science. For several reasons I hold that these pronouncements on method have no special standing as representing Marx's actual theory of science: It is a commonplace in the history of science that working scientists misconstrue the logical or methodological character of their own research,[1] and Marx sometimes overstates the philosophical significance of various elements of his system. (As will be argued in the final section of this chapter, his willingness to "coquette" with a Hegelian manner of expression is an especially regrettable symptom of this difficulty.) Consequently there is a serious risk that these explicit pronouncements will misrepresent the theory of science that actually guided Marx's research.

Nonetheless, several methodological ideas Marx explicitly defends demand discussion—either because of their inherent interest or because of their misuse by both critics and supporters of Marx. Three related methodological ideas will be considered in this chapter. The first will be Marx's often-noted "essentialism": his view of the nature of the social system and the form of scientific explanation appropriate to it. Next we will turn to Marx's "method of abstraction," the set of methodological ideas he presents as an alternative to the experimental methods of the natural sciences. And finally we will consider whether Marx uses a dialectical method in his investigation of capitalism. These ideas are related: Essentialism maintains that the "real" properties of the social system are obscured by its observable features; the method of abstraction is the tool by which Marx proposes to penetrate the misleading appearances of the capitalist system;

and many critics of Marx have believed that both essentialism and the theory of abstraction derive from Hegelian dialectics.

We will find that Marx's methodological writings represent a theory of science almost any contemporary philosopher of science would accept without comment. This theory emphasizes the importance of hypothesis in scientific explanation and provides needed flexibility in relation to empirical evidence. It therefore affords a useful antidote to the uncritical empiricism that often inhibits hypothesis formation in the social sciences. The power of Marx's research program resides precisely in his attempt to construct an explanatory analysis of the capitalist mode of production and not merely a description of its various parts. Thus essentialism has a core of progressive and scientifically legitimate ideas that are perfectly sound according to current views of the nature of science.[2] At the same time it must be acknowledged that Marx often uses a Hegelian terminology in various contexts, but I will argue that these occurrences are superficial and can be omitted from Marx's system without remainder. In particular, I will argue that Marx's method of inquiry is not in any important sense "dialectical."

Essentialism as a Theory of Science

It has often been remarked that Marx possesses an "essentialist" theory of social structure, according to which the real causes of social phenomena are concealed by misleading appearances.[3] Thus G. A. Cohen writes, "Marx frequently pronounced his teaching on essence and appearance when he was at work on *Capital*, which he conceived as an attempt to lay bare the reality underlying and controlling the appearance of capitalist relations of production." [4] Roman Rosdolsky writes in a similar vein, "It is of course true that in *Capital* . . . the 'real inner movement' of capitalist production is constantly contrasted with its 'apparent movement displayed in competition.' And similarly the Hegelian distinction between 'essence' and 'appearance' is consistently employed."[5] And Althusser and Balibar describe Marx's essentialism in similar terms in *Reading Capital*: "For Marx, the science of political economy . . . depends on this reduction of the phenomenon to the essence, or . . . of the apparent movement to the real movement."[6]

Marx's essentialism amounts to the view that the aim of social science is to show how the observable characteristics of the system are shaped by its "inner physiology." This essentialism emerges clearly in Marx's metaphor drawn from astronomy:

> This much is clear: a scientific analysis of competition is not possible, before we have a conception of the inner nature of capital, just as the apparent motions of the heavenly bodies are not intelligible to any but him, who

is acquainted with their real motions, motions which are not directly perceptible by the senses. (*Capital II*, p. 433)

And in the final pages of volume III we find a clear statement of essentialism as a theory of science:

All science would be superfluous if the form of appearance of things directly coincided with their essence. (*Capital III*, p. 956)

Thus science consists in discovering the hidden nature that gives rise to observable phenomena.

This point lies at the heart of most of Marx's criticisms of orthodox political economy. Smith is faulted for failing to clearly distinguish between the inner physiology of the capitalist structure and its visible structure; Ricardo is faulted for not being persistent enough in working out the logic of his own theory of the inner physiology; and vulgar political economy is faulted for avoiding such theories altogether.[7]

[Political economy] has abandoned . . . any kind of solid foundation for a scientific approach so as to be able to retain those distinctions which obtrude themselves on the phenomenal level. This confusion on the part of the theorists shows better than anything else how the practical capitalist, imprisoned in the competitive struggle and in no way penetrating the phenomena it exhibits, cannot but be completely incapable of recognizing, behind the semblance, the inner essence and the inner form of this process. (*Capital III*, p. 269)

This conception also guides the development of Marx's own analysis: He provisionally disregards the complex tangle of observable properties of the capitalist system in order to lay bare the hidden underlying processes that regulate it.

The Ontological Significance of Essentialism

Marx's distinction between visible structure and obscure structure emerges especially clearly from his discussion of Smith's method of political economy in *Theories of Surplus Value*. There Marx observes:

Smith moves with great naivete in a perpetual contradiction. . . . On the one hand he traces the intrinsic connection existing between the economic categories or the obscure structure of the bourgeois economic structure. On the other, he simultaneously sets forth the connection as it appears in the phenomena of competition and thus as it presents itself to the unscientific observer just as to him who is actually involved in the process of bourgeois production. One of these conceptions fathoms the inner connection, the physiology of the bourgeois system, whereas the other takes the external phenomena of life, as they seem and appear and merely describes, catalogues, recounts and arranges them under formal definitions. (*TSV II*, p. 165)

Here Marx is distinguishing between "the inner connection, the physiology of the bourgeois system," and "the external phenomena of life, as they seem and appear." Other phrases he uses to characterize the deep structure of the social system include "the obscure structure of the bourgeois economic structure" and the "intrinsic connection [among economic categories]"; phrases he uses for the observable features of the system include "the external phenomena of life, as they seem and appear" and "the connection as it appears in the phenomena of competition."

As this passage makes plain, essentialism rests on a particular conception of the nature of society: Society consists of an ordered hierarchy of structures, ranging from the observable regularities of the capitalist system — its surface structure — to the hidden principles defining the basic social relations that underlie that observable structure (the "inner physiology"). The social system is the law-governed product of a set of social processes that are not directly observable but that are nonetheless explanatorily fundamental. These underlying mechanisms constitute what Marx calls the *obscure structure* or *inner physiology* of the social formation. Moreover he maintains that as a class system capitalism tends to create appearances that systematically mislead the participant about the nature of the system. Marx refers to this dynamic property of capitalism as "fetishism."[8] These two points — that the defining properties of capitalism are not directly observable and that they are vigorously concealed from the participant through fetishizing institutions — mean that explanation requires abstract hypotheses about those unobservable factors. Sciences content to merely catalogue observable regularities are unlikely to provide adequate explanations of social phenomena.

The inner physiology of capitalism is its system of production: the institutions and structures through which the process is controlled and performed. The visible structure on the other hand is the economy as a system of exchange: the market system of competitive buying and selling of commodities. The essential features of the capitalist system are the relations of production that define it. These relations give rise to an institutional logic that colors the rest of the system. Essentialism thus represents Marx's judgment that (1) the conditions of the system of production *determine* other economic and noneconomic phenomena, and (2) only the relations of exchange are directly identified in the concepts used by the participants within the system. This is discussed clearly in the introduction to the *Grundrisse*, though not with the essentialist language discussed here. "The structure of distribution is completely determined by the structure of production" (*Grundrisse*, p. 95). Political economists have tended to focus their attention on the superficial relations of exchange rather than the essential relations of production. This choice results in their accounts being unsatisfactory as explanations.

Does this mean that the visible structure is illusory or "mere appearance"? The most reasonable position for Marx to take is that the visible structure is real insofar as it is an accurate description of the social system in the vocabulary of observation. The inner physiology is also real to the extent that it represents an abstract account that correctly posits elements of the underlying structure. Both the descriptive and the explanatory accounts represent legitimate levels of analysis, and both are at least potentially true. This view is central to Marx's treatment of the identity of production, distribution, consumption, and exchange in the *Grundrisse*: "The conclusion we reach is not that production, distribution, exchange and consumption are identical, but that they all form the members of a totality, distinctions within a unity. Production predominates not only over itself, in the antithetical definition of production, but over the other moments as well. The process always returns to production to begin anew" (*Grundrisse*, p. 98). Here Marx is acknowledging that every economy contains systems of distribution, exchange, and production, and every economy serves the needs of consumption; thus all aspects of this system are necessary. But the productive system is fundamental in a precise sense: It is the set of institutions that ultimately governs the character of other spheres of the economy. The abstract analysis of the inner physiology is to be preferred, therefore, only because it is more explanatorily adequate. It singles out the influences that establish the nature of the social system and that determine its most fundamental properties. Marx's essentialism sometimes suggests that the underlying level is more real than the visible level, but the point can be put more straightforwardly by maintaining only that the theoretical account makes it possible to explain other aspects of the system, whereas an account restricting itself to a description of the visible structure does not.

On this view, then, the inner physiology of capitalism is the system of production characterized by specific relations of production. But the conditions of exchange tend to mask these relations. Marx recognizes that capitalism is a market system regulated by the conditions of competition. Nonetheless he requires that we abstract from these conditions and analyze the logic of the process of production unmodified by the conditions. Competition is a superficial factor that can be taken account of only after relatively complete consideration has been given to the underlying system of production.

> Competition generally, this essential locomotive force of the bourgeois economy, does not establish its laws, but is rather their executor. Unlimited competition is therefore not the presupposition for the truth of the economic laws, but rather the consequence—the form of appearance in which their necessity realizes itself. (*Grundrisse*, p. 552)

This abstraction leads the resulting analysis to contradict the visible properties of capitalism; for one thing, it implies differences in the rate of profit between industries. This cannot occur in a perfectly competitive economy, so it is necessary to qualify the account at some point. But Marx maintains that these modifications must be introduced at the proper time. They must not obscure the workings of the more basic factors whose institutional implications impose their shape on the whole of the economy.

Essentialism as an Explanatory Paradigm

This conception establishes Marx's basic paradigm: To explain a social phenomenon is to identify the underlying social relations that gave rise to it. The obscure structure represents the most fundamental aspect of the capitalist mode of production, but its effects on the level of surface structure are mediate in the extreme. The methodological principle of abstraction therefore directs social scientists to isolate the most fundamental features and abstract from surface phenomena. (In Marx's view, this means that we must abstract from the phenomena of competition and isolate the logic of capitalist property relations.)

The preceding account of the distinction between obscure structure and visible structure relies heavily on the ordinary scientific distinction between description and explanation, or between surface phenomena and their underlying mechanisms. Smith describes the observable structure of capitalism and makes a preliminary effort to explain it, but this latter effort is unsatisfactory. The visible structure of the social system is the system as it appears to its participants, whereas the obscure structure is the underlying set of processes that define the "essence" of the system, and whose workings determine the details of the visible structure. Moreover, the passage offers a distinctive explanatory paradigm. To explain the observable features of capitalism, it is necessary to isolate the "essential" features of the system from other factors, develop that account in substantial detail, and finally, work out its implications for the observable features of the system.

This point brings in its train a second: the conviction that scientific explanation requires a process of free hypothesis formation. It is not possible to discover the forms of causal influence that exist between elements of social structure through simple observation; rather, it is necessary to construct explanatory hypotheses of these causal relations and then evaluate the hypotheses empirically. Thus essentialism represents a rejection of simple inductivism and naive empiricism. It is not possible to arrive at explanations through simple inductive generalizations; hypotheses are required that are in no way entailed by the phenomena. And too strict an empirical constraint on hypotheses will result in the inability to arrive at any satisfactory model or hypothesis. Therefore scientists must be free to frame hypotheses that are to some degree sheltered from requirements of empirical adequacy.

The methodological principle Marx draws from this conception of the inner physiology of society is that social scientists must seek out the defining social relations of the social system, without being misled by the superficial appearance of the visible structure. To provide an adequate and perspicuous analysis of capitalism it is necessary to abstract from the misleading appearances and examine the inner structure of the system. It is the task of scientific political economy to reveal this inner nature of capitalist economic processes and to show how it expresses itself in the visible structure of capitalism; the phenomena of competition are appearances to be explained, not fundamental explanatory categories themselves.

> [The sphere of circulation] is the sphere of competition, which is subject to accident in each individual case; i.e. where the inner law that prevails through the accidents and governs them is visible only when the accidents are combined in large numbers, so that it remains invisible and incomprehensible to the individual agents of production themselves. (*Capital III*, p. 967)

Classical political economy was ultimately unsuccessful because it failed to recognize the need to formulate an account of the underlying social relations of the capitalist economic structure—the inner physiology. In insisting on this point, Marx is rejecting a very narrow empiricism, since he directs scientists to go beyond the empirical given; so his theory of science will be unwelcome to social scientists who imagine the role of their science to be the systematic description of observable phenomena. This narrow inductivism has been found inadequate for natural science, though, and it is unreasonable to force such a restriction on social science unless we are given some compelling reason.

It should be noted that Marx's distinction between the visible and the obscure structure of a social system is not identical with empiricists' distinction between observation and theory. The "inner physiology" is defined in terms of the notion of explanatory primacy, not that of inherent unobservability. Marx believes that the concepts used by participants in the economy are generally superficial and descriptive, and for the most part reflect incorrect assumptions about the explanatory relations among features of the economy. Moreover he believes that the vulgar political economists typically take over these concepts without analysis or criticism. "The vulgar economist does nothing more than translate the peculiar notions of the competition-enslaved capitalist into an ostensibly more theoretical and generalized language, and attempt to demonstrate the validity of these notions" (*Capital III*, p. 338). But the task of constructing a more adequate set of concepts is not one of formulating highly theoretical postulates; it involves discovering the features of the economy that are in fact explanatorily primary. These features are typically observable, but they are obscured by a plethora of potentially influential factors that are also present. Therefore the task of isolating the essential features requires careful analysis of the sort described in chapter 1.

Thus to identify the property system as a central feature of the inner physiology of the capitalist economy is to assert that this fact is explanatorily fundamental and that the fetishizing and ideological properties of the capitalist order tend to conceal this fact. But it remains true that the details of the property system can be discovered through straightforward empirical investigation, without substantial theoretical assumptions; and explanation amounts to demonstrating (through analysis of the "logic of institutions" created by the property system) the ways in which aspects of the property system impose constraints on the superficial features – the surface structure. The property system is no more hypothetical than the system of market pricing of commodities, but it is explanatorily more central than the latter. Thus essentialism does not appear to be inconsistent with the view taken in chapter 1 that Marx's chief concepts are nontheoretical.

It is apparent from these remarks that essentialism represents an explanatory paradigm for Marx in at least two senses. First, it demonstrates his commitment to a science that explains the phenomena of capitalism rather than merely describing them. And second, it postulates a model of explanation: To explain a phenomenon, it is necessary to discover the institutions or mechanisms that give rise to it, the laws governing these mechanisms, and the processes through which those mechanisms give rise to these phenomena. Thus essentialism directs scientists to seek out particular sorts of explanations (explanations of visible factors in terms of underlying social relations). In the case of capitalism, this means identifying the "essential" social institutions (social relations of production) that underlie the phenomena of capitalism.

Taken together, these elements of essentialism are fairly innocuous portions of a theory of science, given contemporary views in the philosophy of science. Essentialism is highly consonant with the dominant contemporary view of modern science: the conviction that scientific explanation requires more than simple enumeration of factual circumstances. At the same time it must be emphasized that these ideas are of strictly limited use in coordinating scientific research. None provides guidance in the conduct of research or the formation of hypotheses. And this is as we might expect; scientific research rather than scientific method is the substance of science. Vindicating the core assumptions of essentialism at the level of scientific method takes us not one step toward a vindication of Marx's science, and these assumptions are uncontroversial enough to pass as commonplaces today. So the core of essentialism can be regarded as a sound but unremarkable theory of science. It bears emphasis only because many social scientists do *not* accept it and instead remain in the grip of positivist, behaviorist, inductivist ideas that (as Marx writes about other ideas in *The Eighteenth Brumaire*) "weigh like a nightmare on the minds of the living" (*SE*, p. 146). (Consider the sterility, for example, of behaviorism, that in large part derives from its rejection of the need for theoretical hypotheses.)[9] The central ideas of essentialism are necessary but far from sufficient conditions for scientific research; rejecting any of them would seriously reduce the scientific promise of a plan of research, but accepting them does not guarantee results.

Fetishism and Ideology

So far we have treated essentialism as a general view of the nature of scientific explanation. This interpretation of the doctrine puts Marx's system into continuity with much of the history of Western science. A second feature of Marx's view of the inner physiology is more distinctive, however. For Marx maintains that the separation between the appearance and the inner nature of capitalism is especially difficult to bridge because economic appearances positively conceal and mystify the inner nature of capitalism.[10] Capitalism is a class system, and as such it tends to create appearances that systematically mislead participants about the more fundamental social relations. The inner nature of capitalism is a set of relations of production through which one class is dominates and exploits another. If these relations were fully visible, the exploited class would overthrow the exploiters. Stability therefore requires that the social system conceal this reality. This is the scientific significance of the discussion of the fetishism of commodities:

> The mysterious character of the commodity-form consists therefore simply in the fact that the commodity reflects the social characteristics of men's own labour as objective characteristics of the products of labour themselves, as the socio-natural properties of these things. (*Capital II*, pp. 164–165)

> The determination of the magnitude of value by labour-time is therefore a secret hidden under the apparent movements in the relative values of commodities. (*Capital II*, p. 168)

The commodity form systematically conceals the basic social relations of production—wage labor and capital. Descriptions in the vocabulary of participants are therefore distorted; they emphasize the equality and independence of persons and fail to recognize the social relations that regulate them.[11] Consequently accounts of capitalism based on purely observational categories are inadequate; they depend substantially illusory judgments. Science consists in discovering and articulating those fundamental social relations, and this requires disregarding many of the apparent features of the system.[12]

This point is similar to one Marx makes in his discussion of ideology. Consider this statement from *The German Ideology*:

> The fact is . . . that definite individuals who are productively active . . . enter into these definite social and political relations. . . . The production of ideas, of conceptions, of consciousness, is at first directly interwoven with the material activity and the material intercourse of men, the language of real life. The same applies to mental production as expressed in the language of politics, laws, morality, religion, metaphysics, etc. of a people. Men are the producers of their conceptions, ideas, etc. – real, active men, as they are conditioned by a definite development of their productive forces and of the intercourse corresponding to these. (*GI*, pp. 46, 47)

Here Marx is distinguishing between the reality of a set of social relations and the conceptions participants form to represent these relations to themselves and others. These conceptions embody various forms of ideological distortion: "The ideas of the ruling class are in every epoch the ruling ideas, i.e. the class which is the ruling *material* force of society, is at the same time its ruling *intellectual* force. . . . The ruling ideas are nothing more than the ideal expression of the dominant material relationships, the dominant material relationships grasped as ideas" (*GI*, p. 64). That is, Marx holds that the nature of a class system gives the economically dominant class both the opportunity and the incentive to shape the ideas through which men and women conceive of their society. Further, through structural constraints on ideological institutions – universities, newspapers, political organizations, and so on – the class system limits the degree to which social consciousness may correspond to actual social relations.[13] Therefore social scientists and historians must struggle against these conceptions and find ways of uncovering those relations. Thus social science is faced with a problem that does not arise in natural science: namely, that the ideas in terms of which common sense conceptualizes social phenomena are themselves distorted by the workings of a class system and its need to conceal exploitative relationships. By the same token, these passages show that Marx is not an ideological determinist; he does believe that social scientists can expose the real underlying social relations and strip away their ideological trappings through critical empirical research.

The Abstractive Method

Essentialism is accompanied by a distinctive conception of scientific method for social science: the abstractive method. Marx contrasts his method of investigation with that of the natural sciences in the preface to *Capital*.

> The value form . . . is very simple and slight in content. Nevertheless, the human mind has sought in vain for more than 2,000 years to get to the bottom of it, while on the other hand there has been at least an approximation to a successful analysis of forms which are much richer in content and more complex. Why? Because the complete body is easier to study than its cells. Moreover, in the analysis of economic forms neither microscopes nor chemical reagents are of assistance. The power of abstraction must replace both. (*Capital II*, p. 90)

Here Marx is posing a puzzle: How can we explain the fact that political economists from Aristotle through Smith have given relatively accurate descriptions of complex economic phenomena, but they have been unable to discover the simple categories that underlie and explain these phenomena? His answer is that the description of observable phenomena is much easier than the discovery of the causes; furthermore, the means by which we discover underlying causes in the natural sciences—the extension of the senses through microscopes and other instruments or the analysis of substances through chemical or physical tests—are unavailable to us. In place of these analytical techniques, we have only the power of abstraction. The abstractive method requires that investigators formulate abstract hypotheses about underlying mechanisms without restricting their attention to the level of observation.

The Method

The abstractive method is described in the introduction to the *Grundrisse*:

> If I were to begin with the population, this would be a chaotic conception of the whole, and I would then, by means of further determination, move analytically toward ever more simple concepts, from the imagined concrete toward ever thinner abstractions until I had arrived at the simplest determinations. From there the journey would have to be retraced until I had finally arrived at the population again, but this time not as the chaotic conception of the whole, but as a rich totality of many determinations and relations. (p. 100)

This passage provides the basis for a preliminary account of the abstractive method. First, Marx insists that social scientists cannot begin with the fully determinate empirical object—the population. This is the range of observable characteristics of the social formation, but these phenomena are too complex and too misleading to admit of scientific representation. Rather, scientists must form an "abstract" hypothesis about the supporting mechanisms. By a process of "abstraction" scientists formulate a conception of the simplest "determinations"; they then develops these conceptions ("retrace the journey") back through successively more detailed stages to the empirically given whole, the population. But now the population is not a bewildering array of unconceptualized properties but a "rich totality of many determinations and relations."

The abstractive method is needed because capitalist society is a complex whole consisting of social relations, empirical regularities, and social and economic structures. To provide a scientific understanding of that complex range of empirical detail, it is necessary to abstract from observation and arrive at a set of hypotheses about the social relations behind that "chaotic given." Having arrived at the most abstract hypothesis, it is possible to reconstruct, through ever more concretely detailed representations, the complex social formation as it is empirically given; but this depiction of society is no longer chaotic, since it represents a theoretical analysis of the order of the relations that obtain between the empirically given factors. Central to this abstractive method is the view that (1) beneath the empirically given social formation are layers of structure that, taken as a whole, explain the observable shape of that empirical social formation; and (2) the scientifically accurate explanation of the given must reproduce these levels in their correct articulation and order.

According to this methodological principle, then, social scientists should disregard much of the observable structure of the social formation and formulate an abstract hypothesis about the underlying social relations. The result of this method, as Marx well recognizes, will be a descriptively inaccurate account, but Marx maintains that only through such an abstractive analysis that is it possible to arrive at genuine explanations of social phenomena. Having articulated a hypothesis concerning the most fundamental mechanisms of the social system, scientists can then proceed to introduce less fundamental factors—in an orderly way—until at last the theory is in accord with experience; but this process of adding detail is subordinate to the primary abstraction.[14]

This process might seem to be a step away from reality into a false abstraction whose only purpose is pragmatic; Marx's view, however, is just the reverse. He regards the bewildering array of surface phenomena as a falsification of the real nature of the social system; and the abstractive method is the means of penetrating these misleading appearances. Thus abstraction, far from taking us further from the truth, brings us nearer to it. This is its second and more important purpose for Marx: to provide the means of stripping away the inessential and perplexing detail that obscures the operation of the social system, and leave visible only the pure inner workings. The method of abstraction is therefore intimately connected to the essentialism discussed in the previous section. It requires that social scientists *explain* social phenomena on the basis of hypotheses about unobservable social mechanisms (the inner physiology), and it presupposes that the unobservable factors are essential to a satisfactory explanation of the working of the system.

It is this second purpose of the abstractive method that is most important, according to Marx.

> We are deliberately putting forward this law [of the falling rate of profit] before depicting the decomposition of profit into various categories which have become mutually autonomous. The independence of this presentation from the division of profit into various portions, which accrue to different categories of persons, shows from the start how the law in its generality is independent of that division. . . .

> Since we have not investigated up till now the various components into which profit is divided, so that these do not exist for us as yet, the following point is anticipated here simply for the sake of avoiding any misunderstandings. (*Capital III*, p. 320–21)

These passages show that Marx's abstraction is deliberately layered in a way that is significant: It shows the theoretical relationship between the various elements to be considered. In this case, profit is shown to be conceptually prior to the divisions of profit among capitalists (although it is these divisions that we can observe). The significance of this layering arises from the second purpose of abstraction noted earlier. For if our only intention were to pare away part of the picture so that we could see the rest more clearly, we would have a great deal of latitude in the way we proceed; a good many alternative abstractions would suit this objective equally well. Thus Marx might have begun either with a discussion of profit or with one of surplus value; either approach would have reduced the amount of detail necessary to consider. But since Marx's objective is to reveal the inner workings of the social system, his examination must be done in such a way as to abstract from more superficial factors and leave the controlling factors for later analysis. The abstraction itself ought to parallel the relations between factors; it should be designed to show the underlying structure of the social system. Thus in the preceding passage Marx provides a clear illustration of the layered nature of his abstractive system: He takes care to establish the law of the tendency of the rate of profit to fall before a treatment of the more particular forms of profit, exactly because he wants to establish that the law is more fundamental than the subsequent division of profits between capitalists and rentiers. By abstracting from these various forms of profit, Marx abstracts from the differences between the different elements of the capitalist class and points up their much stronger similarity: They all owe their share of the product to the surplus value extracted from the immediate producer.

Once the basic mechanism of the system is laid out systematically, social scientists can proceed to fill in the detail: to take note of the more superficial factors abstracted from initially, to qualify the action of the basis due to factors like competition. "From [the simplest abstractions] the journey would have to be retraced until I had finally arrived at the population again, but this time not as the chaotic conception of the whole, but as a rich totality of many determinations and relations" (*Grundrisse*, p. 100). The outcome of abstractive analysis is once again the whole, but now it is a structured whole: an account that locates each factor in its place within the system and demonstrates the relationships between factors. Abstraction gives us a layered treatment of the social formation, marking factors in terms of the degree to which they are explanatorily fundamental.

Marx's view of the importance of abstraction and hypothesis formation is clear from his critique of Ricardo.[15] He begins by giving Ricardo qualified approval for his method of analysis. Ricardo was the first political economist to distinguish clearly between the problem of describing the visible structure of capitalism and that of providing a theory of its inner nature. Ricardo built on Smith by advancing an abstract theory to explain and unify the various observable phenomena of capitalism: the labor theory of value. Marx faults Ricardo, however, for his failure to stick with his theoretical insights in the face of recalcitrant phenomena.

Ricardo develops a model of capitalism based on the labor theory of value. He then notes that an economy regulated by the principle of value would show different rates of profit in different industries (a circumstance impossible in a competitive market economy), and he consequently modifies the value hypothesis to avoid this consequence. Marx faults this readiness to alter the abstract hypothesis prematurely:

> Instead of postulating this general rate of profit, Ricardo should rather have examined in how far its existence is in fact consistent with the determination of value by labour-time, and he would have found that instead of being consistent with it, prima facie, it contradicts it, and that its existence would therefore have to be explained through a number of intermediary stages. (*TSV II*, p. 174)

Instead of rushing to modify the theoretical hypothesis, Ricardo ought to have developed the theory of value in detail, introducing only in their proper places factors that interfere with its workings. In other words, he ought to have tolerated empirical anomaly in order to develop the theory of value adequately. The contradictions between appearance and theory are to be accounted for only by a methodical and abstractive construction of a theory that introduces new factors step by step. Whereas Ricardo merely postulates an agreement between theory and appearance, Marx argues that social theory must be led through successive levels of abstraction before the contradictions between the two levels can be overcome.

> Science consists precisely in demonstrating how the law of value asserts itself. So that if one wanted at the very beginning to "explain" all the

phenomena which seemingly contradict that law, one would have to present the science before science. (*Correspondence*, p. 196)

We can draw a methodological principle from this critique: Social science should proceed in an orderly and abstractive fashion to construct a theory of the social system that begins with the most fundamental categories and principles, and successively fill in this theory with more superficial factors. The scientifically correct theory of the mode of production must take the form of a layered analysis in which the factors that influence and determine its nature are introduced in an order that reflects their explanatory importance.

This view maintains that social scientists must approach the given social formation with a highly selective eye, disregarding phenomena with little systematic significance and focusing on phenomena that give some indication of the underlying mechanisms. Their objective—initially at least—is not descriptive accuracy; they are not charged with providing a model of the social system that predicts the exact constellation of observable phenomena actually found. Rather, they are expected to provide a model faithful to the hidden nature of the social system, on that isolates its chief characteristics—even if the operation of less fundamental factors obscures the effects of these features. Once this basic model is articulated and developed, qualifications and additional influences can be added to improve the descriptive accuracy of the hypothesis, but this is a strictly secondary enterprise. The chief theoretical work is to reduce the complexity of the given social system to its basis, and this means disregarding the inessential.

Categories and Social Relations

Companion to Marx's theory of abstraction is his view of the nature of the concepts appropriate to scientific political economy. For a central part of the problem of arriving at a perspicuous abstract hypothesis about the inherent character of a social system of constructing appropriate concepts in terms of which to formulate such a hypothesis. In fact, the central distinction between appearance and essence characteristic of essentialism generally can be applied to the nature of scientific concepts: Some concepts serve only to identify the phenomenal features of the social system, and others identify its essential elements. It is Marx's view that the latter concepts must be formulated in such a way as to identify the abstract social relations that define a given social system, and his central criticism of vulgar political economy is that its analysis is couched in the concepts of simple observation. "Vulgar economics actually does nothing more than interpret, systematize and turn into apologetics the notions of agents trapped within bourgeois relations of production" (*Capital III*, p. 956).

Marx's critique in *The Poverty of Philosophy* of Proudhon's political economy provides a clear statement of Marx's view of the nature of economic categories. Several important features emerge from this analysis. First, Marx holds that economic categories express or denote social relations—not material entities, psychological states, or general facts about human nature. "Economic categories are only the theoretical expressions, the abstractions of the social relations of production" (*PP*, p. 109). Prima facie, one might expect that the concepts of economic theory refer to a wide range of possible entities: general factors such as need, material wealth, energy; psychological properties such as desire, aversion, utility; and so forth. Marx argues, however, that the fundamental economic categories refer to none of these, but rather to social relations of production. This view complements the essentialist ideas outlined earlier. Given that the aim of scientific political economy is to provide an analysis of the fundamental properties of a given mode of production, we are naturally led to the conclusion that the concepts of political economy ought to reflect the "atoms" of the mode of production: the social relations of production. This thesis therefore has the force of insisting that political economy is a social science: Economic theory cannot be separated from the social institutions within which economic activity takes place, and there is no distinct and independent range of purely economic phenomena, entities, processes, or the like.

This point emerges quite clearly from Marx's view of the nature of the concept of capital. In the *Grundrisse* he writes ironically in response to the classical economists' idea that capital is simply an accumulated fund of instruments of production: "*Capital* is, among other things, also an instrument of production, also objectified, past labor. Therefore capital is a general, eternal relation of nature; that is, if I leave out just the specific quality that alone makes 'instrument of production' and 'stored up labor' into capital" (*Grundrisse*, p. 86). And what distinguishes capital from other forms of instruments of production? Marx answers this quite clearly in volume III of *Capital*: " . . . capital is not a thing, it is a definite social relation of production pertaining to a particular historical social formation, which simply takes the form of a thing and gives this thing a specific social character. . . . It is the means of production monopolized by a particular section of society" (p. 953). Thus "capital" refers to the social relations through which material instruments of production are used, controlled, and enjoyed—the relations of private property in means of production. (This circumstance accounts for Marx's view that the concepts of capital and wage labor are internally related; each refers to a different side of the same set of social relations of production.)[16] A similar point is made in *Wage Labour and Capital*: "A Negro is a Negro. He only becomes a slave in certain relations. A cotton-spinning jenny is a machine for spinning cotton. It becomes capital only in certain relations" (*WLC*, p. 28). Thus Marx holds quite consistently that the fundamental categories of political economy refer to specific social relations of production.

A second important feature of the concepts of scientific political economy has to do with their historicity. Different modes of production have different "inner natures" or fundamental sets of social relations of production; consequently, political economy must formulate different categories in terms of which to analyze different types of economic structure. In *The Poverty of Philosophy* and elsewhere Marx attacks classical and vulgar political economy for the assumption that the concepts of political economy—on their account, land, labor, and productive wealth—are timeless and equally applicable to all stages of economic life. "Economists express the relations of bourgeois production, the division of labour, credit, money, etc., as fixed, immutable, eternal categories" (*PP*, p. 104). On this approach, it is reasonable to analyze Roman or Greek slavery in terms of the concepts of political economy constructed for modern capitalism,[17] but Marx holds that such an approach necessarily omits the crucial differences between a slave economy and a wage-labor economy.

Marx recognizes that all forms of economic organization must have some features in common. "*Production in general* is an abstraction, but a rational abstraction in so far as it really brings out and fixes the common element and thus saves us repetition" (*Grundrisse*, p. 85). But he holds that these features cannot be explanatorily fundamental precisely because they are in common and consequently cannot account for the crucial organizational differences between distinct modes of production. Rather, "just those things which determine their development, i.e. the elements which are not general and common, must be separated out from the determinations valid for production as such" (*Grundrisse*, p. 85). And the distinguishing features between modes of production may be found only at the level of the defining social relations of production through which productive activity is controlled—the property relations that distinguish slavery from capitalism, or feudalism from slavery. In all economies labor is expended, raw materials and land are transformed, and instruments of production are used; but the crucial differences between modes of production concern the social relations within which these material factors are utilized and controlled.

In insisting on this point about the historicity of economic categories, Marx is making an important methodological statement, but he is also offering the basis of an ideological criticism of orthodox political economy.

> When the economists say that present-day relations—the relations of bourgeois production—are natural, they imply that these are the relations in which wealth is created and productive forces developed in conformity with the laws of nature. These relations therefore are themselves natural laws independent of the influence of time. They are eternal laws which must always govern society. Thus there has been history, but there is no longer any. (*PP*, p. 121)

Marx holds, that is, that the effort to represent capitalist economic categories as timeless and universal serves an ideological function; it supports the idea that capitalism is a natural order and consequently not subject to revolutionary change. "The aim is, rather, to present production . . . as encased in eternal natural laws independent of history, at which opportunity bourgeois relations are then quietly smuggled in as the inviolable natural laws on which society in the abstract is founded" (*Grundrisse*, p. 87).

Against this supposed timelessness, Marx holds that to explain capitalism it is necessary to have a refined and specific analysis of the particular social relations that distinguish it from other modes of production: wage labor and capital. Only once these relations have been singled out and examined will it be possible to work out the institutional consequences of the underlying property forms for the economic structure as a whole.

Marx returns to the logic of concepts in a later methodological work, his *Notes on Adolph Wagner* (1879–80). There he reiterates his criticisms of the ahistorical thinking of orthodox political economists and their efforts to conceive of their science as proceeding from a universal conception of "human nature." Against this view, Marx repeats his own position (clearly articulated in the *Grundrisse* as well, pp. 83–85) that nothing of substance follows from hollow abstractions concerning "man in general." "If it is man situated in any form of society . . . then the determinate character of this social man is to be brought forward as the starting point, i.e. the determinate character of the existing community in which he lives, since production here, hence his process of securing life, already has some kind of social character" (*TM*, p. 189). Thus it is possible to reason only about historically specific circumstances of "man in society" – not about human nature in general. "Individuals producing in society – hence socially determined individual production – is, of course, the point of departure" (*Grundrisse*, p. 83). This is a now-familiar criticism by Marx against political economy; it expresses his insistence that substantive social science can only proceed on the basis of historically specific concepts and knowledge.

More original in the *Notes on Adolph Wagner* is a series of remarks on the origin of scientific concepts. Here the problem is this: How do we arrive at the appropriate set of concepts for political economy? Marx attacks Wagner for holding that the concept of exchange value is the foundation of Marx's system.

> In the first place I do not start out from 'concepts', hence I do not start out from 'the concept of value', and do not have 'to divide' these in any way. What I start out from is the simplest social form in which the labour-product is presented in contemporary society, and this is the 'commodity'. I analyse it, and right from the beginning, *in the form in which it appears*. Here I find that it is, on the one hand, in its natural form, a useful thing, alias a use-value; on the other hand, it is a bearer of exchange-value. (*TM*, p. 198)

Here Marx's point seems to be this: He does not begin his investigation of capitalism with a preconceived set of analytic concepts (e.g., use value and exchange value). Rather, he begins with a view of the fundamental social form of that economic structure—the commodity form—and he chooses concepts that permit him to analyze that form appropriately. Thus the concepts of political economy, according to Marx, cannot be a priori; they must be constructed in close connection with the empirical circumstances they are to describe. It is misleading for Marx to portray this process as not beginning with concepts, since he plainly begins with the concept of the commodity form. Rather, his point is that he begins with a historically and empirically concrete concept, and then constructs more abstract concepts in terms of which to look at and explain this concrete social formation. This feature distinguishes Marx's method both from that of Hegel and that of orthodox political economy. Both Hegel and the political economists begin their theorizing with an a priori set of analytic concepts to use in characterizing their subject matter, whereas Marx holds here that this procedure must be reversed. It is necessary to begin with an empirically detailed social formation and to construct concepts in such a way as to illuminate this specific ensemble of social relations.[18]

In this same passage Marx reiterates the abstractive character of good concepts in political economy. "If we have to analyse the 'commodity'—the simplest economic concretum—we have to withhold all relationships which have nothing to do with the present object of analysis" (*TM*, p. 199). Marx takes this to show that it is methodologically correct in analyzing the commodity form to abstract from the use value of the commodity in order to emphasize the exchange value—because the fact that goods are produced for exchange distinguishes commodity production from other economic forms. This is not to say that use value is unimportant; Marx recognizes full well that a good must have use value to have exchange value. It is the requirements of exchange, however, that creates the characteristic dynamic of capitalist production, and consequently it is this aspect of the commodity form that must occupy the central focus of analysis.

The Organization of a Finished Theory

Having provided an account of the inner physiology of the social system—a model detailing the basic abstract social relations of production—how are we to relate the model to empirical reality? Here Marx introduces a third element to his theory of abstraction. He argues that the analysis of the obscure structure of the social system at which scientists arrive through abstraction can be used to explain the visible structure only through an orderly series of stages of concretization, in which each stage takes account of more superficial factors not previously represented. Factors are to be introduced in order of systemic significance, and sufficient stages are to be included that the ensemble of stages jointly reconstructs the totality—the visible structure as a whole.

As soon as these individual moments had been more or less firmly established and abstracted, there began the economic systems, which ascended from the simple relations, such as labour, division of labour, need, exchange value, to the level of the state, exchange between nations and the world market. The latter is obviously the scientifically correct method. The concrete is concrete because it is the concentration of many determinations, hence unity of the diverse. (*Grundrisse*, p. 101)

The concrete totality, that is, must be reproduced in a theoretically correct order. Marx's most fundamental criticism of Ricardo reflects this view; it arises from Ricardo's treatment of the empirical fact of equal rates of profit across industries. Ricardo modified the theory of value to account for this fact; Marx argued that it must be accounted for "through a series of intermediate stages" (*TSV II*, p. 174). The correct procedure, then, is this: To explain the empirical details of the social system, it is necessary to construct a series of stages of descriptions ranging from the abstract depiction of the inner physiology, through ever more concrete factors, so as to reflect the importance of those factors.

Note that the abstractive method implies a form of explanation that is very distant from the hypothetico-deductive model of neopositivist philosophy of science. It offers an account in which independent factors are introduced in an orderly way. This is in contrast to the ideal of deducing the phenomena from a simple unified theoretical system; the various factors to which such accounts appeal are discovered empirically, not merely through development of a theoretical system.

The organization of *Capital* reflects this principle. Volume I abstracts from the conditions of competition, circulation, and realization in order to characterize the ideal workings of the production process as regulated by value. Volumes II and III are then intended to methodically represent these less fundamental factors, until at last the system correctly describes the empirical reality of capitalism (laws of profit, price, etc.). Thus in volume I Marx defines his purposes as follows:

What I have to examine in this work is the capitalist mode of production, and the relations of production and forms of intercourse that correspond to it. (*Capital II*, p. 90)

He describes his purposes for volumes II and III in these terms:

The second volume of this work will deal with the process of the circulation of capital (Book II) and the various forms of the process of capital in its totality (Book III). (*Capital II*, p. 93)

And in his summary of his results in the opening pages of volume III he writes:

The configurations of capital, as developed in this volume, thus approach step by step the form in which they appear on the surface of society, in the action of different capitals on one another, i.e. in competition, and in the everyday consciousness of the agents of production themselves. (*Capital III*, p. 117)

Thus the circle closes: Volume III is intended to bring the analysis into relation to the full empirical detail of the capitalist system as observed by participants.

This view of the nature of social scientific research and hypothesis formation is mirrored in the detailed organization of Marx's analysis in *Capital*. *Capital* represents a layered theory of the capitalist mode of production that extends from the simplest social relations defining capitalism — wage labor, capital, and the commodity form — to more complex institutional constraints and features of the capitalist system. Thus Marx begins with the commodity form and a discussion of wage labor and capital in volume I; he then describes the process of production in industrial capitalism (volume I); in volume II he provides greater empirical detail by considering constraints imposed by the process of circulation; in volume III he turns to an account of the division of profits among different segments of the bourgeoisie; and he projected, though never wrote, a study of the role of the state in the capitalist economy. The structure of this analysis conforms to the requirements of the abstractive method: Marx begins with the most fundamental but abstract features of the system and works out their institutional logic; he then introduces an orderly series of complicating factors in order to achieve the level of empirical detail that will permit an empirical evaluation of the account as a whole. In the language of the previously quoted *Grundrisse* passage, we arrive at the whole of the capitalist economic structure, "but this time not as the chaotic conception of a whole, but as a rich totality of many determinations and relations" (*Grundrisse*, p. 100), in which the various features of the capitalist economic structure are assigned their proper places within the working of the system.[19]

The Dialectical Method

Those who hold that dialectics are important to understanding Marx's science — Lenin,[20] Lukács,[21] and Ernest Mandel[22] to name several — generally mean that Marx pursues a dialectical method of investigation, that is, that his analysis of capitalist society is guided by principles of dialectical reasoning derived from Hegel's philosophy. Lenin's views are representative of this position.

In his *Capital*, Marx first analyses the simplest, most ordinary and fundamental, most common and everyday relation of bourgeois (commodity) society. . . . In this very simple phenomenon . . . analysis reveals all the contradictions (or the germs of all the contradictions) of modern society. The subsequent exposition shows us the development (both growth and movement) of these contradictions and of this society in the summation of its individual parts, from its beginning to its end.[23]

On this account, once having identified the "central contradiction" of capitalist society we can derive the growth and development of modern society through conceptual or logical analysis of the concepts of capital, labor, commodity, and so on.

This view represents a serious obstacle to the scientific standing of Marx's system, however, because Hegel's dialectical logic is an a priori method of inquiry based on logical analysis rather than on empirical investigation. And in fact critics as diverse as E. P. Thompson,[24] Benedetto Croce,[25] and Sidney Hook[26] have maintained that Marx's use of a dialectical method demonstrates that his reasoning is philosophical rather than scientific. If valid, these criticisms have a good deal of force. In order for Marx to pursue a dialectical method in his study of capitalism, he would have to hold that it is possible to discover the empirical characteristics of the capitalist mode of production on the basis of a priori logical analysis (e.g., of the concepts of capital and the capitalist mode of production). Such a method would exclude Marx's system from standing as a work of science, because it would be a speculative exercise rather than an empirically grounded analysis of historically specific social phenomena. (A current example of just such a "conceptualist" approach to Marxist theory can be found in the writings of Barry Hindess and Paul Hirst.)[27]

It is no doubt true that Marx's mature works contain a certain amount of admittedly Hegelian language and concepts. Marx writes in *Capital*, "I openly avowed myself the pupil of that mighty thinker [Hegel], and even, here and there in the chapter on the theory of value, coquetted with the mode of expression peculiar to him" (*Capital II*, pp. 102–3). And in the same passage he speaks with approval of the dialectical method: "The mystification which the dialectic suffers in Hegel's hands by no means prevents him from being the first to present its general forms of motion in a comprehensive and conscious manner. With him it is standing on its head. It must be inverted, in order to discover the rational kernel within the mystical shell." There is thus some fuel for the argument that *Capital* is not an empirical work but rather a work of materialist philosophy in the Hegelian mode. If these dialectical ideas ran deeply the charge would be compelling. In the following, however, I will argue that Marx is irreconcilably opposed to the use of dialectical logic as a method of inquiry in history or social science. At most the dialectical method represents a highly abstract empirical hypothesis about the nature of social change. I hope therefore to leave the way clear for an interpretation of Marx's scientific method that is in basic agreement with orthodox empirical social science. When Marx goes to work on his detailed treatment of the empirical data of capitalism, he leaves his Hegelian baggage behind.

Dialectics as a Speculative Method

Let us begin with a brief review of some of Hegel's main claims for the scope of his dialectical logic. Hegel believes it is possible to discover the kernel of reason within the chaotic diversity of history through a rigorous development of certain fundamental logical ideas. Thus in his *Encyclopaedia Logic* he writes, "It may be held the highest and final aim of philosophic science to bring about, through the ascertainment of this harmony [between subjective and objective reason], a reconciliation of the self-conscious reason with the reason which is in the world—in other words, with actuality."[28] And in his lectures on the philosophy of history he writes, "The philosophical method may at first strike us as odd, . . . But anyone who does not accept that thought is the sole truth and the highest factor in existence is not in a position to pass any judgement on the philosophical method."[29] Thus the purpose of the philosophy of history is to discover the Reason implicit in the course of history:

> World history . . . represents the development of the Spirit's consciousness of freedom and the consequent realization of that freedom. This development implies a gradual progress, a series of ever more concrete differentiations, as involved in the concept of freedom. The logical, and even more, the dialectical nature of the concept in general, the necessity of its purely abstract self-development, is treated in Logic.[30]

This enterprise is fundamentally an a priori one in which logical or conceptual investigations replace empirical research: "Our aim must be to discern this substance, and to do so, we must bring with us a rational consciousness. . . . Admittedly, philosophy does follow an a priori method in so far as it presupposes the Idea. But the Idea is undoubtedly there, and reason is fully convinced of its presence."[31] Thus a dialectical method appears to be fundamentally *conceptualist* in that it asserts that we can discover features of existing factual circumstances on the basis of logical analysis of certain fundamental concepts (e.g., the concept of Being for Hegel, or the concept of the commodity on Lenin's understanding of *Capital*).

How does Marx respond to these claims for dialectics in the philosophy of history? From the early 1840s on he sharply criticizes the notion of a "philosophical method" that would permit the discovery of substantive truths without the trouble of empirical investigation. Marx's critique of Hegel's *Philosophy of Right* establishes the main themes of his stance toward dialectical reasoning. *The Philosophy of Right* attempts to show the development of the modern state as a rational development conforming to the logical movement of Hegel's *Logic*.[32] In his *Critique of Hegel's 'Philosophy of Right,'* Marx castigates this attempt to deduce the empirical characteristics of a historical phenomenon from the dialectical development of the Idea. For example, early in the *Critique*

Marx considered Hegel's "derivation" of the categories of family and civil society from the Idea of mind. He derides this attempt as a form of "logical, pantheistic mysticism." "Reality is not deemed to be itself but another reality instead. The ordinary empirical world is not governed by its own mind but by a mind alien to it" (*EW*, pp. 61, 62). And the result of this effort to impose Reason on empirical phenomena (i.e., to discover the underlying development of Mind that gives rise to these phenomena) is to produce a spurious semblance of explanation: Appeal to the rational is merely formal, whereas in fact the arbitrary empirical features of the given are simply absorbed into the account without analysis or explanation. "Thus empirical reality is accepted as it is; it is even declared to be rational. However, it is not rational by virtue of its own reason, but because the empirical fact in its empirical existence has a meaning other than itself. The fact that serves as a starting-point is not seen as such but as a mystical result" (*EW*, p. 63).

Marx returns repeatedly to the theme that Hegel's philosophy of history depends on an uncritical positivism; in order to provide Hegel's analysis with any substantive content at all it is necessary for him to smuggle in assumptions about the particular character of the state. Marx applies this point to Hegel's *Phenomenology* in *The Economic and Philosophic Manuscripts*: "In the *Phenomenology* . . . the uncritical positivism and equally uncritical idealism of Hegel's later works, the philosophical dissolution and restoration of the empirical world, is already to be found in latent form" (*EW*, pp. 384–85). But because Hegel's aim is to subsume the particulars of the state under purely rational categories, he cannot provide careful, extensive analysis of this empirical material. "[Hegel] takes an empirical instance of the Prussian or modern state (just as it is – lock, stock and barrel) which can be said to realize this category among others, even though this category may fail to express its specific nature" (*EW*, p. 109; see also pp. 63, 105).

Moreover, Marx argues, Hegel's dialectical "reasoning" is not a form of reasoning at all; it does not transfer logical necessity from premises to conclusions. Thus in discussing Hegel's attempt to assimilate the elements of the state to the model of an organic system, Marx writes, "The use of the word 'hence' creates the illusion of logical rigour, of deduction and the development of an argument. But we should rather ask: why 'hence'? The fact that 'the different members of the organism of the state' are 'the various powers with their functions and their spheres of action' is an empirical fact; but if so, how can this lead us to the philosophical predicate that they are the members of an 'organism'?" (*EW*, p. 67). In summing up Hegel's treatment of the character of the state, Marx states, "If we omit the concrete determinations, . . . we find ourselves confronted by a *chapter of the Logic*. . . . Thus the entire *Philosophy of Right* is no more than a parenthesis within the *Logic*" (*EW*, pp. 73–74).

These passages show that Marx fundamentally rejected the method of investigation contained in *The Philosophy of Right*—the dialectical method. Marx holds that Hegel's form of reasoning does not and *cannot* have deductive force given that it attempts to derive substantive empirical and historical conclusions from pure analysis of abstract concepts. Marx contends that philosophical reasoning cannot recapitulate empirical reality; rather, historical explanation must be founded on empirical investigation. Hegel's philosophy of the state, however, depends entirely on a method of philosophical reasoning. It should be emphasized that this is not merely a rejection of Hegel's metaphysics—his view that Reason is more fundamental than Actuality (his idealism). Rather, it is a rejection of the method of inquiry by which Hegel arrives at conclusions about the actual—history, the nature of the modern state, and so forth. Thus these criticisms represent a rejection of the dialectical method of inquiry itself, not merely the philosophical system at which Hegel arrives.

The essentials of this critique of philosophical method also may be found in Marx's other writings of the period. In *The Holy Family* Marx criticizes the conceptualist features of the dialectic: its effort to arrive at substantive knowledge about nature or society through pure conceptual analysis. He gives an extensive example of the paucity of this method of thought.

> If from real apples, pears, strawberries, and almonds I form the general idea "Fruit", if I go further and *imagine* that my abstract idea "Fruit", derived from real fruit, is an entity existing outside me, is indeed the *true* essence of the pear, the apple, etc.; then in the *language of speculative philosophy* I am declaring that "Fruit" is the *"substance"* of the pear, the apple, the almond, etc. . . . But when the philosopher expresses their existence in the speculative way he says something *extraordinary*. He performs a miracle by producing the real natural objects, the apple, the pear, etc., out of the unreal creation of the mind "the Fruit". (*HF*, pp. 72–75)

Marx sees this process of conceptual analysis as wholly vacuous and incapable of leading to substantive knowledge of the nature of particular kinds of fruit; instead, it gives a spurious sense of having subsumed the actual under the categories of reason. And Marx suggests with his example that this method of thought characterizes the Hegelian dialectic in all its applications to concrete phenomena: "Hegel very often gives a real presentation, embracing the thing itself, within the speculative presentation. This real reasoning within the speculative reasoning misleads the reader into considering the speculative reasoning as real and the real as speculative" (*HF*, p. 76). This is the same point Marx makes in referring to Hegel's uncritical positivism: Insofar as there is substantive content in Hegel's analysis, it is tacitly smuggled in and made to look as though it were the consequence of purely logical reasoning.

In *The German Ideology* Marx extends this critique by offering a more specific description of his alternative to dialectical philosophy, a method based on empirical analysis of existing material conditions. Here Marx criticizes the Young Hegelians (Bauer, Strauss, Feuerbach, Stirner, etc.) in the following terms. "German criticism has . . . never quitted the realm of philosophy. Far from examining its general philosophic premises, the whole body of its inquiries has actually sprung from the soil of a definite philosophical system, that of Hegel. Not only in their answers but in their very questions there was a mystification" (*GI*, p. 40). The core of the charge here is that these systems continue to rely on a philosophical method of inquiry, and such a method is inherently a priori and speculative. By contrast, in *The German Ideology* Marx and Engels call for a method grounded in specific material facts, and these facts can be discovered only through concrete empirical investigation. "The premises from which we begin are not arbitrary ones, not dogmas, but real premises from which abstraction can only be made in the imagination. . . . These premises can thus be verified in a purely empirical way" (*GI*, p. 42). And a few pages later: "Empirical observation must in each separate instance bring out empirically, and without any mystification and speculation, the connection of the social and political structure with production" (*GI*, p. 46). Throughout *The German Ideology*, then, Marx and Engels distinguish their method from that of the Young Hegelians on these grounds: Whereas the Young Hegelians continue to pursue a philosophical and speculative method, Marx and Engels pursue a method that is grounded in empirical research into material conditions. It is difficult to read these passages as anything but a renunciation of a dialectical logic as a method of inquiry.

Consider finally Marx's critique of Proudhon's political economy in *The Poverty of Philosophy*. This discussion is particularly important in the present context because Proudhon *does* attempt to base political economy on a dialectical method of inquiry and explanation, and Marx sharply rejects the possibility of such a science. "What Hegel has done for religion, law, etc., M. Proudhon seeks to do for political economy" (*PP*, p. 107). Marx describes Proudhon's method in these terms: "If one finds in logical categories the substance of all things, one imagines one has found in the logical formula of movement the *absolute method*, which not only explains all things, but also implies the movement of things" (*PP*, p. 107). Thus Proudhon's project is defined as an attempt to assimilate the categories of political economy to an abstract logical system derived from Hegel's *Logic*, and then to derive the economic laws that can be deduced from this system. Marx's commentary on this approach makes it plain that he thinks it entirely spurious as a technique of scientific inquiry. Here again Marx's critique of dialectics as a speculative, a priori analytic tool is sharp and unforgiving. "The moment we cease to pursue the historical movement of production relations, of which the categories are but the theoretical expression, the moment we want to see in these categories no more than ideas, spontaneous thoughts, independent of real relations, we are forced to attribute the origin of these thoughts to the movement of pure reason. . . . Or, to speak Greek—we have thesis, antithesis and synthesis" (*PP*, p. 105). "Apply this method to the categories of political economy, and you have the logic and metaphysics of political economy" (*PP*, p. 108).

In opposition to Proudhon's dialectical method of discovering the inner logic of historical developments through analysis of philosophical concepts, Marx describes a thoroughly empirical and historical means of inquiry: "When . . . we ask ourselves why a particular principle was manifested in the eleventh or in the eighteenth century rather than in any other, we are necessarily forced to examine minutely what men were like in the eleventh century, what they were like in the eighteenth, what were their respective needs, their productive forces, their mode of production—in short, what were the relations between man and man which resulted from all these conditions of existence" (*PP*, p. 115). Thus Marx puts forward the concrete empirical method described in *The German Ideology* as his alternative to the speculative method of dialectical logic.

Throughout his writings of the mid-1840s, then, Marx devotes a great deal of attention to the construction of a full-blown critique of dialectical logic as a method of inquiry for historical knowledge or social science. He judges that Hegel's dialectics—in Hegel's own hands and in those of his followers—is necessarily speculative and a priori; it depends on logical analysis rather than empirical investigation; and it must be decisively rejected in favor of concrete empirical research before genuine historical explanation can proceed.

Discussion of Dialectics in *Capital*

It would seem that Marx's criticisms of Hegel's dialectics would preclude his employing a similar philosophical method in his own analysis of the capitalist mode of production. And in fact Marx rarely uses the terminology of the dialectic in his mature writings. The only explicit discussion of the dialectical method in *Capital* occurs in the postface to the second edition, in which the celebrated "inversion" of Hegel's method is described. And analysis of this discussion will bear out the conclusions drawn in the previous section: that Marx endorses an empirical method in explicit opposition to a speculative dialectical method. In this passage Marx organizes his comments around a sympathetic review of *Capital* by a Russian who provided a brief but accurate description of Marx's effort to discover historically specific "laws of motion" for the capitalist mode of production. Marx then comments, "But what else is he depicting but the dialectical method?" (*Capital II*, p. 103). Thus Marx seems to endorse the view that his method is dialectical. Let us try to discover what he means by this comment, however. Marx first distinguishes between the method of presentation and the method of inquiry:

The latter has to appropriate the material in detail, to analyse its different forms of development and to track down their inner connection. Only after this work has been done can the real movement be appropriately presented. If this is done successfully, if the life of the subject-matter is now reflected back in the ideas, then it may appear as if we have before us an *a priori* construction. (*Capital II*, p. 102)

As this passage makes plain, the method of inquiry is empirical — one through which the concrete specifics of the social formation are investigated in detail. It is through this sort of research that investigators can discover the "inner connections," that is, the causal, functional, and institutional connections among elements of the social system. It is at the stage of the presentation of results that the notion of a dialectic enters: The material discovered through this empirical inquiry is reorganized into a coherent, orderly system, and the material is presented so as to reveal the internal dynamic through which social phenomena develop and change. But the apparent apriority of the latter construction is illusory; it has resulted not from an a priori logical analysis but rather from a detailed empirical investigation. Consequently, this passage does not support the idea that Marx pursues a method of investigation grounded in a dialectical logic.

Thus Marx's description of his method here is entirely consistent with the materialist, empirically grounded method that he advocated in *The German Ideology* and elsewhere. This passage requires that investigators immerse themselves in the fullness of the empirical detail of the social formation and "appropriate the material in detail" before attempting to provide explanations of the social processes discovered. There is nothing "dialectical" about this process; on the contrary, it would be more appropriate to describe it as a form of (enlightened) empiricism: not a simple inductivist version of empiricism, but rather a view according to which hypotheses and theories must be evaluated in terms of their compatibility of a wide range of empirical data.[33]

We are now able to interpret Marx's celebrated remark that with Hegel the dialectic is "standing on its head. It must be inverted, in order to discover the rational kernel within the mystical shell" (*Capital II*, p. 103). What is the rational kernel, and what is the mystical shell, of the dialectic? And in what sense does Marx "invert" Hegel's method? It is Marx's endorsement of Hegel's view of the historicity of social institutions and their internal dynamics of change that leads him to speak with favor of the dialectical method. "In its rational form . . . [the dialectical method] regards every historically developed form as being in a fluid state, in motion, and therefore grasps its transient aspect as well" (*Capital II*, p. 103). On my view of Marx's meaning, the rational kernel of Hegel's method is the empirical hypothesis that historical and social processes develop according to an internal dynamic, and that it is possible to provide a rigorous analysis and explanation of historical change based on knowledge of that dynamic. Moreover, Marx plainly accepts Hegel's view that change in history proceeds through substantive contradictions. These theses characterize history as a law-governed process, and one whose changes develop as the result of internal contradictions. Thus the kernel of Hegel's dialectic, on Marx's view, is not methodological at all, but rather a revealing insight into the character of social reality. These theses are empirical hypotheses (albeit formulated at an extremely high-level).

The mystical shell of Hegel's method, by contrast, *is* methodological—and perniciously so. It is Hegel's belief that pure a priori analysis can allow him to discover the key to this internal dynamic. This assumption is the "logical mysticism" identified by Marx in the *Critique of Hegel's "Philosophy of Right,"* and he emphatically rejects this philosophical method. Finally, the inversion Marx proposes requires that instead of beginning with ideas and attempting to reproduce the material world in thought, we must begin with the material world and attempt to arrive at ideas that adequately describe its real characteristics. "With me the reverse is true: The ideal is nothing but the material world reflected in the mind of man, and translated into forms of thought" (*Capital II*, p. 102). Thus Marx's inversion of Hegel's method is materialism; more exactly, it is a form of empiricism because it stipulates that knowledge of the material world may be acquired only through detailed, rigorous investigation of concrete empirical and historical circumstances. And what are the methods appropriate to this sort of investigation? They are the methods of empirical science.[34]

According to this line of thought, the notion of dialectic has application to Marx's system only in connection with his conception of the nature of social processes. Marx's view of social change is that societies develop through contradictions or oppositions: conflict between the forces and relations of production, conflict between classes, or conflict between the particular interests of a single capitalist and the common interests of the capitalist class.[35] This view of society is not a philosophical theory, however, but rather part of an extended empirical analysis of the actual workings of capitalist society. And no esoteric form of "dialectical reasoning" is needed to identify these inner contradictions in capitalist society; rather, these features of society can be investigated through straightforward empirical investigation. The thesis of historicity is not an a priori methodological principle, but rather a substantive empirical hypothesis. Consequently, these passages seem to support the general view being advanced here: that Marx's method of inquiry is solidly empirical and analytic, and the notions of dialectics come in only at the level of general hypotheses about the nature of social change and historicity.

A final point relevant to the issue of the putatively dialectical nature of Marx's system concerns the form of necessity Marx attributes to various kinds of social change—the idea of the "inner logic" of the social system. Marx believes it possible to devise an account of the capitalist mode of production that shows that the system develops according to a necessary pattern. He puts this point quite clearly in the preface to the first edition of *Capital*. "Intrinsically, it is not a question of the higher or lower degree of development of the social antagonisms that spring from the natural laws of capitalist production. It is a question of these laws themselves, of these tendencies winning their way through and working themselves out with iron necessity" (*Capital II*, pp. 90-91). Given its basic institutional structure, that is, a certain pattern of development is fixed. This claim might seem to parallel Hegel's view that certain fundamental historical processes occur as a result of rational necessity—for example, the emergence of the individualism of the Reformation as a stage of freedom.[36] Empirically minded social analysts must reject this sort of claim, so it is necessary to find some other way of construing Marx's use of the notion of "necessity."

Fortunately such an alternative is available in the form of the idea of an institutional logic (discussed in chapter 1). For if we consider Marx's typical explanations in detail, we find that they depend on a kind of analysis that attempts to relate macrolaws (e.g., the laws of motion of capitalism) to the circumstances of individual choice and action that produce these laws in the aggregate. Marx's explanations are founded on examination of the circumstances of choice within which rational individuals deliberate in the context of capitalist economic institutions. Marx identifies a set of motivational factors and constraints on action for a hypothetical capitalist (or proletarian, landlord, etc.). These constraints are represented largely by the property relations that define the economic structure of capitalism. Marx then tries to work out an account of the most rational strategies available to the agent in these circumstances of choice. Finally he attempts to discover the aggregate consequences of large numbers of agents acting rationally in these ways, and the laws of motion that he discovers are just such aggregate results of rational individual behavior. Therefore the form of necessity Marx postulates for his social laws derives from the notion of rational action within constrained circumstances of choice. (This model of explanation will be developed in chapter 5.)

This treatment of Marx's reasoning allows us to give content to his idea that the capitalist mode of production has a logic of development without appealing to a dialectical method. The interpretation offered here establishes the consonance of the idea of a "social logic" with other branches of empirical social science by showing how this idea of a logic of development can be understood in terms of the notion of composing aggregate macrolaws out of analysis of the

actions and strategies of rational individuals within the social relations of capitalism. That is, the necessity identified earlier (e.g., of the laws of accumulation
within capitalism) is not logical necessity but an institutional imperative: Given
the basic institutions of capitalism, the tendencies represented by the laws of accumulation will naturally tend to occur as a result of the actions of large numbers
of rational individuals (capitalists, proletarians, landlords, etc.). And this form
of necessity is fully consistent with the idea of a scientific explanation of social
phenomena. This interpretation therefore further supports the view that Marx's
system does not depend on Hegel's philosophical system or on his dialectical
method.

A close reading of the only passage in *Capital* in which Marx discusses the
dialectic confirms, then, the general view taken here that Marx does not follow
a dialectical method at all; at most he takes over from Hegel a set of highly abstract empirical hypotheses about the character of social change. These hypotheses have much the character of the "research hypotheses" discussed in chapter
2: They guide Marx's research, they possess empirical content, and they are in
no sense a priori truths.

Examination of Marx's explanatory practice throughout his economics further
confirms this conclusion. Marx makes no effort to show that the social system
is guided by a logic of ideas or concepts, and he does not endeavor to reconstruct
the "necessary" process of development of a social formation on the basis of
analysis of abstract logical concepts. He does not, in short, use primarily
philosophical forms of argument. This case cannot be made in detail here; indeed, it is the purpose of much of this book to present the forms of analysis and
argument contained in *Capital*. But we have already seen in preceding chapters
that the tools of inquiry Marx employs in his substantive work in *Capital* are
not derived from esoteric philosophical systems; rather, he relies on conventional tools: causal and functional analysis, analysis of various forms of institutional logic, and mathematical economic analysis.

D.-H. Ruben takes much this view of the dialectic in his useful article "Marxism and Dialectics." There he writes, "There is nothing irreducibly important
about the terminology of the dialectic in Marx; everything to be said dialectically
can be said in other words. . . . Since so many Marxists use dialectical terminology on a par with the magic of 'abracadabra' and related mumbo-jumbo in
order to conjure away problems, it is refreshing to see . . . that one
can . . . wholly dispense with the dialectical vocabulary."[37] Likewise, Terrell
Carver judges that dialectical ideas play only a minor role in Marx's system.
"Marx's debt to Hegel is highly overrated in conventional accounts. . . . Dialectic plays a minor but important role in Marx's method for seeking the anatomy of civil society in political economy, and in no sense represents a key of
his work or a summary of it."[38] The minor role that Carver attributes to the di-

alectic in Marx's system is closely related to that identified earlier: Marx's agreement with Hegel that social processes represent patterns of opposition and need to be understood as such. Finally, in "Dialectic and Ontology"[39] Milton Fisk argues that the idea of dialectics in Marx has a primarily ontological significance rather than a methodological role: It pertains to Marx's view of the nature of social reality rather than to his view of the method by which that reality should be investigated.[40]

Marx's "Galilean" Empiricism

These remarks permit us to attribute a form of empiricism to Marx that distinguishes him both from Hegel's dialectical method and from traditional British empiricism. First, Marx emphatically rejects the notion of an a priori method of analysis through which it might be possible to discover the substantive features of capitalism without empirical research. Thus in discussing the problem of explaining differences between the development of Roman commercial society and that of European capitalism, Marx writes, "Events strikingly analogous but taking place in different historical surroundings led to totally different results. By studying each of these forms of evolution separately and then comparing them one can easily find the clue to this phenomenon, but one will never arrive there by using as one's master key a general historico-philosophical theory, the supreme virtue of which consists in being supra-historical" (*Correspondence*, p. 294). This skepticism about the possibility of philosophical methods of discovery is a feature of Marx's theory of knowledge that brings him into relation to traditional British empiricism. Second, Marx advocates a positive theory of knowledge that gives central place to empirical investigation and the use of factual evidence in the evaluation of scientific claims. Marx expressly asserts that substantive knowledge of history and social institutions can be discovered only through concrete historical and empirical investigation. He thus recognizes that his account of capitalism is ultimately to be assessed in terms of its compatibility with empirical evidence concerning the institutions of capitalism. Marx thus holds—again, parallel to traditional British empiricism—that empirical evidence is the ultimate and irreplaceable foundation of knowledge.

At the same time, these parallels do not establish that Marx's epistemology is simply a development of traditional British empiricism; for Marx also offers stinging criticisms of this tradition (particularly well documented in James Farr's articles on this subject).[41] The chief dimension of these criticisms is Marx's rejection of inductivism or naive empiricism—the view that it is possible to arrive at scientific explanations of natural or social phenomena through simple observation, inductive generalization, and careful record keeping. In particular, Marx rejects the idea that explanatory concepts can be extracted from the observable features of capitalist society. His most serious charges against political economy rotate around this point. "Vulgar economy . . . here as everywhere sticks to appearances in opposition to the law that regulates and explains them" (*Capital II*, pp. 421–22). Similar criticisms are advanced in volume III as well:

> The vulgar economist does nothing more than translate the peculiar notions of the competition-enslaved capitalist into an ostensibly more theoretical and generalized language, and attempt to demonstrate the validity of these notions. (*Capital III*, p. 338)

And in *Theories of Surplus Value* Marx criticizes Adam Smith in much the same terms: "[Smith] sets forth the connection [among the phenomena of capitalism] as it appears in the phenomena of competition and thus as it presents itself to the unscientific observer just as to him who is actually involved in the process of bourgeois production" (*TSV II*, p. 165). Against this family of views, Marx holds that scientific explanation requires abstract hypotheses about the underlying mechanisms that give rise to observable phenomena:

> This much is clear: a scientific analysis of competition is not possible, before we have a conception of the inner nature of capital, just as the apparent motions of the heavenly bodies are not intelligible to any but him, who is acquainted with their real motions, motions which are not directly perceptible by the sense (*Capital II*, p. 433).

Thus a scientific explanation of the phenomena of capitalism must be founded on a hypothesis about their causal structures and mechanisms. This is the feature of Marx's epistemology that entitles us to call it "Galilean" empiricism. Naive empiricism, because it assumes that scientific knowledge consists of simple inductive summaries of past experience, is unacceptable.

This model is a paradigm of explanatory adequacy that can be traced to Galileo: The view that explanation requires hypotheses about the basic "reality" that may seem to contradict the evidence of the senses.[42] Galileo was not content merely to describe the behavior of rolling balls; rather, he undertook to account for their motion in terms of the underlying forces. Richard Westfall discusses Galileo's significance in just these terms:

> Throughout the *Dialogue*, it is Simplicius [the exponent of Aristotelian mechanics] who asserts the sanctity of observation. . . . Inertial motion is an idealized conception incapable of being realized in fact. If we start from experience, we are more apt to end with Aristotle's mechanics, a highly sophisticated analysis of experience. In contrast, Galileo started with the analysis of idealized conditions which experience can never know.[43]

E. A. Burtt puts the point similarly. He observes that Galileo insists on the primacy of empirical evidence in astronomy ("observation is the true mistress of astronomy"), but he also points out that for Galileo "the world of the senses is not its own explanation; as it stands it is an unsolved cipher" that must be explained abstractly.[44]

On the one hand we cannot deny that it is the senses which offer us the world to be explained; on the other we are equally certain that they do not give us the rational order which alone supplies the desired explanation.[45]

Both Burtt and Westfall accept the point central to our concern: For Galileo, to explain an empirical phenomenon, it is necessary to provide an abstract hypothesis concerning its hidden causes.

This Galilean strand of Marx's method led him to an explanatory study of capitalism, but he recognized that this theory must be evaluated in terms of its empirical credentials. Thus Marx insisted that the ultimate criterion of his success would be the degree to which the observable features of the capitalist economy confirmed his analysis, and he attempted to work out the empirical consequences of the theory to permit such an evaluation. Given that theoretical hypotheses cannot be *deduced* from the empirical evidence, the most we can require is that researchers be current with the chief forms of available evidence, and critically evaluate their theories in light of evidence. Marx certainly satisfies that condition. *Capital* is a theoretical examination of the inner nature of capitalism, but this analysis is tempered by ongoing attention to empirical data and by genuine commitment to empirical controls on the finished theory.

Marx's empiricism, then, has a strikingly modern flavor. It rejects both the claims of speculative thought and the excessively narrow limitations of inductivism. In place of both, it holds that science must be formulated in terms of abstract hypotheses that attempt to identify hidden causal mechanisms, and that these hypotheses must be justified in terms of a relevant array of empirical evidence. This position is sharply distinguished both from Hegel's conceptualist method and from crude empiricism. Hegel's system is idealist and a priori; it is not concerned with using empirical evidence to evaluate the adequacy of the account. Marx's system, by contrast, is thoroughly grounded in a careful use of empirical evidence. At the same time, Marx insists, against crude empiricism, that scientific explanations can proceed only on the basis of hypotheses about supporting mechanisms, structures, causes, and the like, that cannot be directly inferred from empirical data.

We can conclude that Marx's system does not depend on any serious commitments to a dialectical method or to esoteric philosophical theories. Rather, his scientific practice and his comments about his "dialectical" method are best construed as a commitment to ordinary forms of investigation and reasoning. Marx's scientific treatment of the capitalist mode of production (CMP) is substantive and largely independent of any of the philosophical views suggested by his occasional "coquetting" with Hegel. Marx's system offers a set of hypotheses concerning the causal structure of the CMP, and it provides rigorous analysis of some of the most important features of the CMP. Marx's model is *not* philosophically troublesome; its chief hypotheses are couched concepts that, though theoretical, are not primarily philosophical: class, property, surplus appropriation. And as will be seen in detail in the final two chapters, Marx uses evidence carefully and appropriately. This usage stands solidly within a sensible commitment to empirical control of belief.

5

Explanation

One of Marx's chief purposes in *Capital* is to *explain* various aspects of capitalist society—the falling tendency in the rate of profit, the formation of a "surplus population" of unemployed workers, and the tendency toward rapid technological innovation within capitalism, to name a few. As we saw in the preceding chapter, Marx distinguishes his account from that of orthodox political economy—in part, at least—on the ground that orthodox political economy is content merely to describe the phenomena of capitalism, whereas Marx is intent on explaining these phenomena. This chapter considers the nature of the explanations that Marx advances. We will begin by reviewing recent arguments about the role of macroexplanations in Marxism. Then we will inquire into Marx's standards: What is the logical structure of an explanation in his system, what sort of reasoning establishes an explanation, and what sorts of regularities does he refer to? We will find Marx's explanations usually derive macrophenomena from abstract descriptions of the microprocesses that lead to those phenomena: the circumstances within which representative class actors pursue their interests within capitalist property relations. Thus Marx's explanatory practice corresponds closely to the "institutional-logic" account introduced in chapter 1.

Microfoundations and Social Explanation

Marxist social science commonly has advanced macroexplanations of social phenomena in which the object of investigation is a large-scale feature of society and the explanans a description of some other set of macrophenomena. Some Marxist social scientists have recently argued, however, that macroexplanations stand in need of *microfoundations*: detailed accounts of the pathways by which macrolevel social patterns come about.[1] These theorists have held that it is necessary to describe the circumstances of individual choice and action that give rise to aggregate patterns if macroexplanations are to be adequate. Thus to explain the policies of the capitalist state it is not sufficient to observe that this state tends to serve capitalist interests; we need an account of the processes through which state policies are shaped and controlled so as to produce this outcome.[2] Let us begin with a brief discussion of several issues that have stimulated the "microfoundations" program.

One important family of criticisms of macro-Marxism involves the status of functional explanations in social science. As we saw in chapter 2, a functional explanation is one that explains a social institution in terms of its beneficial consequences for the social system as a whole, or some important subsystem. Functional explanations have always played a prominent role in Marxist theory, and they have acquired fresh importance through the influential writings of Gerald Cohen.[3] Cohen argues that historical materialism depends crucially on functional explanations. Thus in considering the relationship between the forces and relations of production Cohen writes, "We hold that the character of the forces *functionally* explains the character of the relations. . . . The favoured explanations take this form: *the production relations are of kind R at time t because relations of kind R are suitable to the use and development of the productive forces at t, given the level of development of the latter at t.*" [4] And in discussing the relation between the economic structure and the superstructure, Cohen goes on to say that "the general explanatory thesis is that given property relations have the character they do because of the production relations property relations with that character support."[5]

Consider an example: Why did the American state enact the New Deal legislation of the 1930s that gave rise to various forms of social welfare programs? This is a particular problem for Marxist political theory because such programs appear to conflict with capitalist interests through higher taxation, a less pliable working class, and so forth. Various explanations are possible of this phenomenon; but a classical Marxist explanation goes along these lines: The New Deal was enacted because it contributed to the overall stability of capitalist property relations in a time of serious social and economic crisis.[6] This is a functional explanation. It explains an event (the enactment of the New Deal) in terms of its beneficial contribution to the workings of a larger system (in this case, the capitalist social and economic system).

In an important series of writings, Jon Elster argues that functional explanations are inherently incomplete in social science (though not in biology); they must be supplemented by detailed accounts of the social processes through which the needs of social and economic systems influence other social processes so as to elicit responses that satisfy those needs.[7] The problem is that it is almost always possible to come up with *some* beneficial consequences of a given institution; so to justify the judgment that the institution exists *because* of its beneficial consequences we need to account for the mechanisms that created and reproduced the institution in such a way as to show how the needs of the system as a whole influenced the development of the institution. Thus in discussing Co-hen's functional explanation of procapitalist bias in the media, James Noble writes, "Although I do not doubt that unions get less favorable press than management during a strike, recognizing that fact is not the same thing as ex-plaining it. Clearly, Cohen's proffered explanation *hints* at some mechanism of social control. But until we have an account of that mechanism, we just do not have an explanation. Neither our understanding nor our ability to change the role of the press in capitalist society has been advanced."[8] If a functional relationship is to count as explanatory, therefore, it is necessary to have some idea of the causal mechanisms that establish and preserve the functional relationship.[9]

In explaining the New Deal case, therefore, we must ask through what sorts of mechanisms might the need for economic stability influence the process of legislation? Once we begin looking for such mechanisms, several different sorts of candidates appear. We might observe that legislators and lobbyists have eco-nomic interests; they can perceive that existing economic crisis endangers those interests; and finally, they can undertake legislation designed to damp down the tensions that have been generating social and economic crisis. Alternatively we might imagine a social analogue to the principle of natural selection in biology: Various legislative packages are put forward, and those that best serve the needs of the economic structure are selected over those that do not serve those needs well.[10] But without some account of the mechanisms that give rise to the rela-tionship, the functional explanation is unsatisfactory.

These remarks suggest a moral for macro-Marxism: To explain a phenome-non it is not sufficient to demonstrate that the phenomenon has consequences that are beneficial for the economy, or for the interests of a particular class. Rather, it is necessary to provide a detailed account of the micropathways by which the needs of the economy, or the interests of a powerful class, are imposed on other social phenomena so as to elicit beneficial consequences. Thus the macroexpla-nation is insufficient unless it is accompanied by an analysis at the level of in-dividual activity – a microanalysis – that reveals the mechanisms that give rise to the pattern to be explained.

Consider now a second issue underlying the call for "microfoundations" for Marxian explanations: the gap between the interests of a group as a collective and the interests of the individuals who compose the group. (John Roemer refers to this as the "aggregation gap.")[11] "Rational-action" explanations depend on identifying an individual's interests and then explaining that person's behavior as the rational attempt to best serve those interests. This model is often extended to account for collective behavior of groups as well: To explain the group's ac-tion, it is sufficient to identify the collective interests of the group, and then view the group's behavior as a rational attempt to further those interests. However, Mancur Olson and others have made it plain (as simple analysis would show in any case) that it is not sufficient to refer to collective interests in order to explain individual behavior.[12] As Roemer puts the point, "Just because it would be rational for society as a whole to adopt an innovation does not prove it will be adopted, for that adoption may not be in the interests of the individuals who have the power to prevent its adoption."[13] We all share a common interest in clean air, but none of us has a motive based on rational self-interest in refraining from polluting the air. Nonetheless, some Marxist macroexplanations appear to depend precisely on a conflation of collective and private interests.

This problem is well illustrated in recent discussions of the problem of "revolutionary motivation." Why should Marxist theory expect that workers will contribute to a revolutionary overthrow of the capitalist system? The orthodox Marxist answer is that revolution is in the collective interest of the working class: "The proletarians have nothing to lose but their chains. They have a world to win" (*R1848*, p. 98). Thus Marx's view appears to be that it is in the rational self-interest of the working class as a whole to overthrow capitalism and establish socialism. The unseen difficulty here, however, is that of getting from a circumstance's being in the collective interest of a group to its being in the individual interests of members of the group. In an important recent article Allen Buchanan argues, "Even if revolution is in the best interest of the proletariat, and even if every member of the proletariat realizes that this is so, so far as its members act rationally, this class will not achieve concerted revolutionary action."[14] The problem here is that the collective good to the working class as a whole may be outweighed in each individual case by the cost of participation, with the result that the individual is compelled rationally to refrain from participation.[15]

The point of Buchanan's arguments is not that collective action is impossible; on the contrary, examples are to be found throughout social life: industrial strikes, consumer boycotts, peasant rebellions. These arguments are intended rather to show that collective action cannot be explained merely by alluding to collective interests shared by rational individuals. Thus it is necessary to have a fine-grained account of the variety of motives that lead individuals to act out of regard for collective interests rather than narrowly defined self-interest. Once we begin to look for appropriate microlevel mechanisms that bridge the aggregation gap, we find a variety of possible solutions. First, we discover that successful organizations have mechanisms for translating collective interests into individual interests, for example, through "in-process" benefits to encourage cooperation or sanctions to discourage free riders.[16] Second, we may find that shared moral values within a group play an important causal role in collective action, for example, the values of solidarity, fairness, reciprocity, or unwillingness to be a free rider.[17] And third, we can turn to the findings of social psychologists, political scientists, and others, who have studied motivation within small social groups. Here we find that individuals are subject to a richer variety of factors than a simple model of economic rationality would allow: ties of friendship and kinship, loyalty to the village or political party, religious motivations, or surges of "crowd psychology" (manifested in spontaneous peasant uprisings, for example).[18]

The lesson to be gained from this example is simple but important: In reasoning about the behavior of a group—whether a social class, a labor union, or a group of professional athletes—it is not sufficient to consider only the collective interests of the group. It is also necessary to provide a microanalysis of the motives, interests, and incentives that cause the individuals of the group to act, as well as an account of the various ways in which collective interests both shape and are frustrated by individual interests. A macroexplanation that concludes that a factory will be easily unionized because it is in the plain and evident interest of the workers as a whole fails on this criterion; it does not establish that individual workers will have sufficient motivation to contribute to this public good.

In both cases the objection being advanced to macro-Marxism is grounded in a recognition that there are no supraindividual actors in a society. Classes are not corporate entities that determine their interests collectively and act in concert; states are not rational, deliberative entities standing over and above the persons who occupy office; and technologies and economic systems do not "seek out" social institutions that best suit them. Rather, all social entities act through the individual men and women who occupy roles within them. So if we are concerned to know why a given social entity undergoes a particular process of change, we need to have specific information about the motives and opportunities that define the positions of the persons who make up that entity. Explanations of social phenomena are unsatisfactory unless they identify the microprocesses through which explanandum and explanans are linked.[19] In this way the call for microfoundations represents a variant of methodological individualism.[20] Notice, however, that unlike classical arguments for methodological individualism, this stricture does not rest on a general ontological thesis about the relation between individuals and social entities. Further, it is not merely an application of a general program of reductionism in science, according to which higher-level laws must be grounded in lower-level laws. The microfoundational requirement depends rather on specific problems that arise in two common forms of macroexplanation: functional and collective-interest explanations. One consequence of this point is that there may be legitimate forms of macroexplanations in social science that are not subject to these specific criticisms, and in that case microfoundational arguments would be silent.

Marx's Explanations

Let us now consider some of Marx's explanations in *Capital*. These examples have been chosen because they are fairly representative of Marx's explanatory practices and because they illustrate the most important features of his explanatory paradigm.

First, consider Marx's account of the tendency toward technological innovation and ever-higher productivity in capitalism. Marx observes in *The Communist Manifesto* that capitalism, uniquely among historical modes of production, is characterized by continual revolutions in the means of production driven by the need to increase output.

> The bourgeoisie cannot exist without constantly revolutionizing the instruments of production, and thereby the relations of production, and with them the whole relations of society. . . . The bourgeoisie, during its rule of scarce one hundred years, has created more massive and more colossal productive forces than have all preceding generations together. Subjection of nature's forces to man, machinery, application of chemistry to industry and agriculture, steam navigation, railways, electric telegraphs, clearing of whole continents for cultivation, canalization of rivers, whole populations conjured out of the ground—what earlier century had even a presentiment that such productive forces slumbered in the lap of social labour? (R1848, pp. 70, 72)

This striking historical feature of capitalism demands explanation, and in *Capital* Marx attempts this.

Before it is possible to explain the tendency within capitalism toward ever greater productivity, it is necessary to have a clear definition of the productivity of labor. Marx provides this as follows:

> By an increase in the productivity of labour, we mean an alteration in the labour process of such a kind as to shorten the labour-time socially necessary for the production of a commodity, and to endow a given quantity of labour with the power of producing a greater quantity of use-value. (*Capital I*, p. 431)

There is a tempting macroexplanation of the tendency toward technical innovation: Increasing productivity in wage-goods industries leads to a fall in the wage, that in turn (other things being equal) leads to an increase in the average rate of profit within the capitalist economy as a whole.[21] Consequently one might propose that productivity-enhancing technical innovations occur within the capitalist economy *because* they increase the average rate of profit by cheapening the wage. This is a functional view of the tendency; it explains the macropattern in terms of its beneficial consequences for another macropattern (increase in the average rate of profit). However, this explanation enters the aggregation gap immediately. For the fact that productivity-enhancing innovations would lead to a collective good for the capitalist class does not suffice to demonstrate that these innovations will come into being; rather, it is necessary to demonstrate that the particular circumstances, interests, and opportunities of representative capitalists would lead them to adopt such innovations.[22] In other words, it is necessary to provide a microfoundation for this macroexplanation.

Significantly, Marx's own explanation rests on just such a microanalysis. His own account of technical change occurs in volume I of *Capital* (pp. 429–511). Marx's explication of this tendency is based not on a consideration of the collective interests of the capitalist class but on the fact that *individual* capitalists are rational and goal directed, and their fundamental purpose is to organize their enterprises to create the greatest amount of profit. Marx considers the result for the individual capitalist of introducing a labor-saving technical innovation into the manufacture of shirts.

> The individual value of these articles is now below their social value. . . . The real value of a commodity, however, is not its individual, but its social value. . . . If, therefore, the capitalist who applies the new method sells his commodity at its social value of one shilling, he sells it for 3d. above its individual value, and thus he realizes an extra surplus value of 3d. . . . Hence, quite independently of [whether the commodity is a wage good], there is a motive for each individual capitalist to cheapen his commodities by increasing the productivity of labor. (*Capital I*, p. 435)

The profitability of an enterprise within a competitive market is determined, on the one hand, by the technical characteristics of the process of production employed by the enterprise itself—its intrinsic efficiency or productivity—and on the other hand, by the average productivity of the industry as a whole that is involved in the production of the good. The price of the good is determined by the average productivity of firms in the given industry. Firms producing the good at the average rate of productivity earn the average rate of profit; those producing with less-than-average efficiency earn a lower rate of profit, and those producing with higher-than-average productivity earn a higher rate of profit. Consequently each capitalist has an immediate need for technical innovations that will allow him to produce the good at higher-than-average productivity (or lower-than-average net cost of production). Given this incentive, each capitalist energetically seeks out new production techniques with higher efficiency. The collective result for individual capitalists is a powerful tendency toward technical innovation.

This positive incentive leads also to a fear that equally stimulates the search for new techniques: Each capitalist realizes that his competitors also are vigorously looking for such innovations and that technologically laggard firms will find themselves at a severe competitive disadvantage. Given this double incentive – the prospect of short-term superprofits and fear that a competitor will acquire a lethal advantage through technological innovation – each capitalist actively seeks new techniques with higher productivity. Here, then, Marx has identified the microexplanation of this strong tendency in. Once a new technique becomes widespread, however (as it inevitably will, given competitive pressures), the selling price falls to the new social value of the commodity, the superprofits disappear, and the search for new techniques begins anew. Note that the argument does not assume that it is impossible for a given capitalist to behave differently, but rather that different behavior is irrational and likely to lead to the bankruptcy of the nonconformist firm. So the system as a whole will display the derived tendencies even if some individual capitalists resist.[23]

Here we have Marx's microexplanation of the surge toward ever-greater labor productivity: Every capitalist has a well-defined individual interest in increasing the productivity of his enterprise, in the form of an opportunity to achieve greater-than-average amounts of profit. And this interest is entirely independent from the collective good to be gained through increases in wage-goods industries for the capitalist class as a whole. The individual capitalist is concerned not with social averages (average rate of profit, average cost of production, or the industrywide average amount of labor time expended in the production of a given commodity) but with his own amount and rate of profit.

This account proceeds by demonstrating how the tendency toward higher productivity is related to certain basic features of the capitalist mode of production: the incentives, prohibitions, and opportunities that define the position of the capitalist within the capitalist economy. In chapter 1 we identified the social relations of production (property relations) as basic to Marx's analysis of capitalism. These relations establish the structural and functional characteristics of the capitalist mode of production that Marx regards as essential, and this explanation rests on that account. Marx tries to show that certain systemic properties follow from these underlying social relations: If capitalists act generally so as to maximize the rate of profit, the long-term result will be a tendency toward higher productivity.

We should note that this is not a functional explanation and does not fall into the aggregation gap. Marx does not explain the macrotendency toward technical innovation in terms of the collective benefits this tendency confers on the economy as a whole, or on the capitalist class; instead, he relies on an analysis of the microcircumstances of the representative capitalist and an account of the aggregate consequences that these circumstances have for the development of capitalism as a whole.

Consider next Marx's discussion of the falling rate of profit (already discussed in chapter 3; *Capital III*, pp. 317–75). Marx begins with a simple algebraic illustration of the behavior of the rate of profit under different ratios of constant capital to variable capital (the organic composition of capital, or the capital-labor ratio). For the sake of illustration he assumes that the rate of surplus value (s/v) is 100 percent (that is, half the workday reproduces the value of the wage, and the other half provides the surplus value). And he considers a series of increasing capital-labor ratios, showing that the rate of profit in these cases decreases as the organic composition of capital (c/v) rises.

> The same rate of surplus-value, therefore, and an unchanged level of exploitation of labour, is expressed in a falling rate of profit. . . . If we further assume now that this gradual change in the composition of capital . . . occurs in more or less all spheres . . . and that it therefore involves changes in the average organic composition of the total capital . . . then this gradual growth in the constant capital, in relation to the variable, must necessarily result in a *gradual fall in the general rate of profit*, given that the rate of surplus-value . . . remains the same. . . . Moreover, it has been shown to be a law of the capitalist mode of production that its development does in fact involve a relative decline in the relation of variable capital to constant. (*Capital III*, pp. 317–18)

On these assumptions, the rate of profit may be represented in the following way:

$p' = s / (c + v)$ (rate of profit)
$m = c/v$ (organic composition of capital)
$p' = s' / (1 + m)$ ($s' =$ rate of surplus value$-s/v$)

And it follows algebraically that if m rises while s' remains constant, p' must fall.

This argument is simpler than the explanation for the tendency toward higher productivity just considered. It rests on two substantive premises and an analytic representation of the rate of profit. The assumptions are these:

1. The organic composition of capital tends to rise through the development of capitalism.
2. The rate of surplus value remains relatively constant through the development of capitalism.

The first premise describes a tendency for which Marx offers several independent arguments; the second is an abstractive assumption he knows to be false. (He writes, for example, that "the rate of surplus value, with the level of exploitation of labour remaining the same *or even rising*, is expressed in a steadily fall-ing general rate of profit"; *Capital III*, p. 319.) In addition to these substantive assumptions, Marx also makes an analytic assumption concerning the correct way of representing profits by equating aggregate surplus value with aggregate profit. Marx's explanation of the falling rate of profit, therefore, may be put as follows: The rate of profit tends to fall over time because the organic composi-tion of capital tends to rise through the development of capitalism.

The crucial premise here concerns the behavior of the organic composition of capital over the development of capitalism. In volume I Marx argues that the organic composition will tend to rise.

> Apart from natural conditions . . . the level of the social productivity of labour is expressed in the relative extent of the means of production that one worker, during a given time, . . . turns into products. The mass of means of production with which he functions in this way increases with the productivity of his labour. . . . This change in the technical compo-sition of capital, this growth in the mass of the means of production, as compared with the mass of the labour-power that vivifies them, is reflected in its value-composition by the increase of the constant constituent of capi-tal at the expense of the variable constituent. (*Capital I*, p. 773)

Here Marx's reasoning is fairly straightforward. We have already seen that he establishes a powerful tendency within capitalism toward technical innovations that enhance the productivity of labor. In this passage Marx is making a simple empirical observation: Such innovations tend to involve the replacement of hu-man labor with machinery and more sophisticated technology, that in turn leads to increases in the capital-labor ratio. As Marx puts the point in volume III, "This progressive decline in the variable capital in relation to the constant capi-tal . . . is just another expression for the progressive development of the social productivity of labour, that is shown by the way that the growing use of machin-ery and fixed capital generally enables more raw and ancillary materials to be transformed into products in the same time by the same number of workers" (*Capital III*, p. 318). Thus the organic composition of capital tends to rise as a natural consequence of the quest for productivity-enhancing innovations.

There is a second source of the tendency of the organic composition of capital to rise, however, that is independent of the quest for higher productivity. This concerns the perennial class conflict between capital and labor. Given that the wage is ultimately determined by the relative demand for labor, and given that skilled labor can command higher wages than unskilled, every capitalist has two incentives to choose capital-intensive techniques in order to economize on wage costs. First, he has an incentive to seek out technical innovations that will re-place skilled labor with unskilled. "The capitalist . . . progressively replaces skilled workers by less skilled, mature labour-power by immature, male by female, that of adults by that of young persons or children" (*Capital I*, p. 788). Innovations that permit this sort of replacement will tend to be capital-intensive, however, as they involve the replacement of human skill and strength with automatic machinery and new sources of power. Second, the capitalist has an incentive to choose innovations that will reduce the demand for labor altogether by replacing living labor with machinery. "The mechanism of capitalist production takes care that the absolute increase of capital is not accompanied by a corresponding rise in the general demand for labour. . . . Capital acts on both sides at once. If its accumulation on the one hand increases the demand for labour, it increases on the other the supply of workers by 'setting them free' [i.e., displacing them with machinery]" (*Capital I*, p. 793). Technical innovations guided by these incentives may be neutral from the point of view of the productivity of labor, but they are desirable to the capitalist because they strengthen his bargaining position within the labor market.

Let us turn now to Marx's second substantive assumption in this explanation. This concerns the behavior of the rate of surplus value; for the sake of argument Marx supposes that the rate of surplus value tends to remain constant. This assumption is more problematic; on the contrary, on Marx's own suppositions it is likely that the rate of surplus value will rise as productivity rises. This increase is likely because of the following fact: In times of rising productivity, commodities fall in value because they embody less labor time. Consequently the contents of the wage basket fall in value, and therefore the value of the wage falls as well. The rate of surplus value could remain constant only if the wage basket is increased in size—that is, by raising the worker's real standard of living. And this increase would mean that capital and labor share proportionally in the increases in productivity. Given Marx's theory of the wage, however, according to which the absolute level of the wage is determined by the relative bargaining strengths of the capitalist and the workers, it would seem probable that the capitalist can insist on a more favorable division of the fruits of increasing productivity than strict proportionality.

The most likely outcome, therefore, in times of rising productivity, seems to be this: Workers will negotiate a wage that permits some increase in the real standard of living, but the net value of the new wage basket will be less than the value of the old wage basket. And in fact Marx seems to endorse this conclusion: "Under the conditions of accumulation [when the productivity of labor is rising] . . . a larger part of the worker's own surplus product . . . comes back to them in the shape of means of payment, so that they can extend the circle of their enjoyments, make additions to their consumption fund of clothes, furniture, etc., and lay by a small reserve fund of money. But these things no more abolish the exploitation of the wage-labourer, . . . than do better clothing, food and treatment, . . . in the case of the slave" (*Capital I*, p. 769). (This position is often described as a theory of "relative immiseration"; that is, workers improve their position absolutely but decline relative to the capitalist.)[24] If this argument follows from Marx's premises, however, Marx must predict that the rate of surplus value will rise in conditions of increasing productivity. But if the rate of surplus value rises, the behavior of the rate of profit is indeterminate unless we can conclude that the increase in the rate of surplus value is less than that needed to offset the increase in the organic composition of capital.

Marx does not consider this problem in detail in connection with the falling rate of profit. He does, however, acknowledge in these same passages that the rate of surplus value may well rise during periods of increased productivity. "The law of the falling rate of profit, as expressing the same *or even a rising rate* of surplus-value, means in other words: taking any particular quantity of average social capital . . . an ever greater portion of this is represented by means of dead labour and an ever lesser portion by living labour" (*Capital III*, p. 322). Generally his position appears to be that the rate of surplus value will rise, but the organic composition of capital will rise more rapidly, with the result that the rate of profit will continue to fall.

This argument has many faults,[25] but it offers a clear example of Marx's explanatory paradigm. To account for some phenomenon P, it suffices to show that P follows from the ideal account of the capitalist mode of production (presented in chapter 1) through economic analysis. The argument rests on substantive institutional assumptions (in particular, those aspects of the Marx's account of capitalism that are involved in his argument for the tendency toward higher productivity), and on the technique by which Marx analyzes profits (the labor theory of value). Thus this argument too represents an attempt to show that some particular feature of the capitalist mode of production follows from the basic structural and functional properties that define it.

Let us turn now to a third example: Marx's explanation in volume I of *Capital* of the formation of a pool of unemployed workers. Marx takes it as an empirical fact that capitalism creates such a fund of workers (that he refers to as the "industrial reserve army") even in times of prosperity, and the empirical evidence of the past century has tended to confirm this judgment. A functional explanation of this is even more attractive in this case than in the previous case. For the existence of a pool of unemployed workers obviously benefits the capitalist class: It serves to depress the wage, since employed workers face competition for jobs from unemployed workers. It also reduces the militancy of the working class for much the same reason: It reduces their bargaining position. One might hold, therefore, that the industrial reserve army exists within capitalism *because* it confers these benefits on the economy.[26]

However, this is not Marx's strategy in attempting to explain this phenomenon. Instead, he considers the incentives and opportunities that define the position of the representative capitalist in order to discover what leads to a relative surplus of workers. (*Capital I*, pp. 781–94) He holds that each capitalist has an incentive to reduce his own labor costs, and this incentive becomes stronger as the wage rises. He also has the incentive to reduce his dependence on the available fund of workers in order to improve his own bargaining position. Moreover, the capitalist has the opportunity to discover labor-saving innovations (through research and development), and he has the power to introduce these innovations (through his rights as property owner). Therefore the representative capitalist will tend to effectively use such labor-saving innovations. The collective consequence of this individual tendency, however, is that demand for labor falls periodically; in other words, there is a tendency in capitalism toward the creation of a pool of unemployed and underemployed workers. Once again we find that Marx avoids available functional explanations, preferring instead to account for the macrophenomenon in question in terms of the circumstances of choice within which the representative capitalist acts.

We now turn to a final example that is rather different from the first several. This is Marx's analysis of the conditions of reproduction that regulate the capitalist economy. In volume II of *Capital* Marx writes,

What we were dealing with in both Parts One and Two [of volume II] . . . was always no more than an individual capital, the movement of an autonomous part of the social capital. . . . However, the circuits of individual capitals are interlinked, they presuppose one another and condition one another, and it is precisely by being interlinked in this way that they constitute the movement of the total social capital. Just as, in the case of simple commodity circulation, the overall metamorphosis of a single commodity appeared as but one term in the series of metamorphoses of the commodity world as a whole, now the metamorphosis of the individual capital appears as one term in the series of metamorphoses of the social capital. . . . What we now have to consider is the circulation process of the individual capitals as components of the total social capital. . . . Taken in its entirety, this circulation process is a form of the reproduction process. (*Capital II*, pp. 429–30)

This passage appears to suggest, first, that the account up to this point is based on analysis of the conditions within which individual capitalists make choices, and second, that some properties of the capitalist economy emerge only at a macrolevel (the level of "social capital"). Thus Marx appears to accept that much of his examination takes the form of reasoning about individuals and thus conforms to the requirements of "microfoundational Marxism." But he also seems to hold that others of his explanations are inherently macrolevel explanations and are not reducible to microlevel processes or mechanisms.

A particularly clear candidate for such an explanation is Marx's model of simple reproduction. The problem that defines this area may be formulated as follows:

> How is the capital consumed in production replaced in its value out of the annual product, and how is the movement of this replacement intertwined with the consumption of surplus-value by the capitalists and of the wages by the workers? (*Capital II*, p. 469)

That is, what are the conditions of equilibrium that must be satisfied if the capitalist production process is to reproduce itself from one circuit of capital to the next? Marx's chief contribution to this problem is contained in his "two-department" model of reproduction (*Capital II*, pp. 468–509). He breaks down the capitalist economy into two sectors: capital-goods industries and consumption-goods industries. Consumption demand is equal to the wages paid and surplus generated in both sectors; capital-goods demand is equal to the sum of the capital consumed in each sector. Therefore if the two sectors are to be in equilibrium – that is, if all commodities produced in both sectors are to find buyers in the marketplace – the total value of the goods produced in Department I (production goods) and Department II (consumption goods) must satisfy the following condition of equilibrium (derived in chapter 3):

$$c_{II} = v_I + s_I \ (\textit{Capital II}, \text{ p. 478}).$$

If production should violate these conditions of equilibrium, various forms of "crises of realization" will occur: There will be an excess of either capital goods or consumption goods that cannot find buyer in the marketplace.

This analysis plays an important role in Marx's economics; like Quesnay's *Tableau economique* (or Leontief's input-output models), it provides Marx with an analytic device for describing the flow of commodities through the capitalist economy from industry to consumption to capital-goods reproduction. We can reasonably ask, however, what this scheme *explains*. It appears to be no more than a scheme of description in terms of which to abstractly characterize the flow of commodities in the capitalist economy. It might be thought that this scheme gives Marx a basis on which to account for certain types of economic crisis in the capitalist economy: To explain why realization crises occurred in a particular time period, Marx might say that the conditions of smooth reproduction were not satisfied in both sectors. However, as Marx himself observes, "It is a pure tautology to say that crises are provoked by a lack of effective demand or effective consumption" (*Capital II*, p. 486). In fact, these conditions of equilibrium are a clear example of what Thomas Schelling calls the "inescapable mathematics of musical chairs"[27] – a result that is logically inescapable given a particular way of formulating a problem. For once we choose to describe the capitalist production process in terms of labor values, these equilibrium relations follow tautologically. Consequently this scheme by itself does not explain anything. In order to account for *either* the occurrence of crisis or an extended period of smooth reproduction, it is necessary to consider the mechanisms at work within the economy that either disrupt or preserve the requirements of equilibrium.

Marx's Explanatory Paradigm

Let us now consider what model of explanation appears to underlie these examples. Several of these examples suggest that Marx's explanations commonly involve two types of analysis, each of which is compatible with the requirement that macroexplanations be provided with microfoundations. First, Marx's accounts depend on an examination of the circumstances of choice of rational individuals. Marx identifies a set of motivational factors and constraints on action for a hypothetical capitalist and then tries to determine the most rational strategies available to the capitalist in these circumstances of choice. This effort falls within what would be called today the theory of rational choice.[28] In the example of technical innovation Marx isolates the dominant motive guiding the hypothetical capitalist, namely, the desire to increase the amount of surplus value realized within his unit of production. He then canvasses available means for increasing the amount of surplus value, for example, intensification of the labor process and productivity-enhancing technical innovation. And he holds that the capitalist, as a rational agent with the specified motives, will arrive at one such strategy and organize his unit of production accordingly: He will seek out technical innovations that reduce the cost of production.

This sort of analysis turns on identifying the motives that guide the hypothetical capitalist, and the institutional constraints within which the capitalist acts. (It might be possible to increase the rate of profit by enslaving the workers, but this is institutionally prohibited.) And this analysis arrives at a prediction about the choices to be made by the hypothetical capitalist. The centrality of these sorts of motivational factors in Marx's explanations conforms to the view of his system offered in chapter 1. There we found that the social relations of production (the property relations) constitute a central part of Marx's account of capitalism. These social relations constitute both Marx's theory of the interests that guide each class and his account of the institutional prohibitions and incentives that constrain the choice of strategies for each class. Thus Marx's view of the social relations of production of capitalism is explanatorily fundamental; it identifies the central factors relevant to analyzing capitalist behavior.

Crucial to the adequacy of this sort of account is the accuracy of the premises of the analysis: the account of the motives and interests of the hypothetical actor, and of the institutional context of prohibitions and incentives within which the actor deliberates. Significantly, much of Marx's effort in *Capital* goes into articulating these premises in detail. Throughout we find Marx identifying the capitalist in terms of the incentives and constraints that define his position within the economic structure: "The valorization of value . . . is [the capitalist's] subjective purpose, and it is only in so far as the appropriation of ever more wealth in the abstract is the sole driving force behind his operations that he functions as a capitalist, i.e. as capital personified and endowed with consciousness and a will" (*Capital I*, p. 254). And in discussing the capitalist's domination of the worker within the process of production Marx writes, "*Capital* developed within the production process until it acquired command over labour . . . , in other words the worker himself. The capitalist, who is capital personified, now takes care that the worker does his work regularly and with the proper degree of intensity" (*Capital I*, p. 424). Throughout, then, Marx is concerned to identify the specific incentives, constraints, and opportunities that define the activities of the representative capitalist. And these assumptions about interests, motives, and constraints are central to his explanations of macrocharacteristics of the capitalist economy.

Although this sort of analysis is couched in terms of hypothetical individuals, Marx makes it plain that it is a way of reasoning about behavior of individuals as members of classes. "Individuals are dealt with here only in so far as they are the personifications of economic categories" (*Capital I*, p. 92). Such an account thus functions as a form of microclass analysis—not of how classes as wholes behave but rather of how representative members of particular classes behave.[29] Naturally information about the microlevel of class behavior has consequences for how the class behaves collectively as well. Moreover Marx makes it plain that the circumstances of choice are fundamentally different for members of different classes. The interests and opportunities of capitalists are fundamentally different from those of workers. Consequently an important function of ongoing capitalist institutions is to maintain the fixity of these class distinctions, that is, to prevent workers from disturbing the opportunities for serving capitalist interests defined by capitalist property relations. "The capitalist process of production, therefore, seen as a total, connected process, i.e. a process of reproduction, produces not only commodities, not only surplus-value, but it also produces and reproduces the capital-relation itself; on the one hand the capitalist, on the other the wage-labourer" (*Capital I*, p. 724).

A second part of this model of explanation involves an attempt to determine the consequences for the system as a whole of the forms of activity attributed to the hypothetical capitalist at the preceding stage of analysis. That is, every capitalist is subject to essentially the same motives and is under roughly the same institutional constraints. On these assumptions, what sorts of collective patterns are likely to emerge as a result of a large number of actors making the choices attributed to the hypothetical actor? In some cases this analysis is quite simple. If every capitalist is striving to discover productivity-enhancing technical innovations, then the collective result is increases productivity of labor in the econ-omy as a whole. And this increase leads to a fall in the value of the wage, which in turn leads to an increase in the rate of surplus value. Other collective conse-quences are less obvious. For example, each individual capitalist introduces technical innovations in order to increase his firm's rate of profit. The collective consequence of this incentive, however, is a fall in the average rate of profit (due to the attendant rise in the organic composition of capital). This form of analysis is closely related to modern collective-choice theory:[30] What are the collective consequences of a particular structure of individual action?

Such accounts can be made more fine grained by considering some of the varieties to be found both among motives or interests and among institutional environments. Thus Marx distinguishes between the patterns of activity charac-teristic of finance capital and of industrial capital. "Merchant's or trading capital is divided into two forms or subspecies, commercial capital and money-dealing capital, that we shall go on to distinguish in such detail as is needed in order to analyze capital in its basic inner structure. And this is all the more necessary in so far as modern economics, and even its best representatives, lump trading capital and industrial capital directly together and in fact completely overlook trading capital's characteristic peculiarities" (*Capital III*, p. 379). "[Competition] reproduces a new financial aristocracy, a new kind of parasite in the guise of company promoters, speculators and merely nominal directors" (*Capital III*, p. 569). These different segments of the capitalist class occupy somewhat dissimi-lar positions within the system of property relations, and this gives rise to varied interests and motives for these "subspecies" of capitalists. Likewise, small and large capitalists occupy substantially different environments, for instance, in terms of their access to sources of credit; consequently the strategies appropriate to serving even similar interests often will be different. Given these disparities of interests and opportunities between groups of capitalists (merchant, finance, and industrial capital; large and small capital; domestic and international producers; etc.), a full treatment of the behavior of capitalism requires separate analysis of representative individuals from each group.

These explanations display a third important feature of Marx's paradigm: his use of economic analysis. Many of Marx's explanations are founded on a piece of economic analysis grounded in the labor theory of value. This is most clearly illustrated in his account of the falling rate of profit. Like the previous examples, this account depends on institutional assumptions of the sort discussed earlier (e.g., the incentive each capitalist has toward maximizing the rate of profit). But it also depends on the particulars of Marx's economic model of capitalism (the theories of value and surplus value and his theory of profit). In order to make out an argument concerning the probable movement of the rate of profit through the development of capitalism, it is necessary first to have a theory of profit:

What factors determine or influence the rate of profit? Such a theory will identify a set of factors that determine the rate of profit. On Marx's account, aggregate profits for an enterprise are equal to aggregate surplus value created. "Profit is . . . a transformed form of surplus-value, a form in which its origin and the secret of its existence are veiled and obliterated. In point of fact, profit is the form of appearance of surplus-value, and the latter can be sifted out from the former only by analysis" (*Capital III*, p. 139). (It should be noted, however, that this is an abstractive assumption; Marx recognizes that the movement from consideration of a single representative firm to the collective description of a capitalist economy as a whole has radical consequences. In particular, differences among industries of the organic composition of capital entail that profits will not equal aggregate surplus within given enterprises in competitive conditions.)

Having put forward a theory of profit in terms of surplus value, it is possible for Marx to represent various aspects of the capitalist economy in terms of a set of algebraic formulas. Thus the rate of profit may be represented as

$$p' = s / (c + v) \; (Capital \; III, \text{ p. 133}),$$

where s is surplus value, c equals constant capital, and v is variable capital. This representation of the rate of profit permits Marx to investigate mathematically the relationship between the rate of profit and changes in other factors (e.g., the rate of surplus value or the organic composition of capital). It is thus possible to use techniques of mathematical analysis to investigate the relationships between a variety of different economic quantities.

In order to discover the behavior of these economic quantities, however, Marx is forced to return to the sort of analysis described earlier. His argument concerning the behavior of the organic composition of capital (the capital-labor ratio), for example, depends on the incentives and opportunities available to the capitalist. Thus Marx's use of economic analysis supplements rather than replaces the aforementioned pattern of investigation. Having established the mathematical relations that hold between the rate of profit and the organic composition of capital, Marx's explanation resumes the pattern described earlier: It attempts to determine the institutional consequences of a pattern of representative capitalist behavior.

Marx's Explanatory Scheme

This discussion suggests that Marx typically puts forward an explanation that may be represented as involving the following sorts of propositions.

1. All participants (capitalists, workers, landlords, etc.) are rational actors in that (a) they have a clear conception of what their interests are,

and (b) they form rational beliefs about the environment in which they pursue their interests and the strategies that are available to them.

2. The social relations of production of capitalism establish specific motivational factors – incentives and prohibitions – for the capitalist through the system of property relations, for example, (a) achieve profits; (b) accumulate capital; (c) establish competitive advantage over other capitalists; (d) refrain from fraud and violent coercion; and (e) respect private property.

3. Specific features of capitalism define the environment of choice within which the capitalist acts: (a) technical features of capitalist production (e.g., the introduction of steam power); (b) opportunities for international trade; (c) shifting balance of forces between social classes (e.g., rising power of working class organizations); (d) opportunities for influencing state actions; (e) availability of finance and credit; and so on.

4. Given a particular set of incentives, prohibitions, and opportunities, the representative capitalist will arrive at a particular strategy S.

5. Given that S is the best strategy for the representative capitalist, and given that all capitalists are rational, all capitalists will adopt S.

6. When S is universally adopted, it has the consequences C in the aggregate (e.g., it leads to a rising organic composition of capital, a falling rate of profit, the creation of an industrial reserve army, or a tendency toward concentration of capital).

The first three propositions represent the general content of Marx's typical explanations. They identify the background circumstances within capitalist society that define the basic organization and development of capitalism, according to Marx. Propositions (2) and (3) summarize the full detail of the analysis throughout *Capital* of the fundamental features of the capitalist economic structure. These propositions represent the substantive empirical content of Marx's accounts in that they identify the features of capitalism that on Marx's view determine various aspects of capitalist phenomena. Moreover, as formulated here propositions (2) and (3) are blank checks; in particular explanations Marx draws freely on general facts about the capitalist economic structure as discovered throughout his research. Consequently it is not possible to represent briefly the forms of knowledge that may be brought to bear in this position in the typical argument; rather, this position represents the open-ended sort of descriptive analysis seen throughout *Capital*. Proposition (1) represents the assumption of rationality necessary in order to bring to bear on actual capitalist behavior reasoning about rational individual choices. (Obviously this is an idealized assumption; real capitalists will display a range of degrees of rationality.)

The final three propositions describe the specific arguments and analysis contained in a given explanation. They represent Marx's effort to reason from general features of capitalist economy to specific features of capitalist society in need of explanation (e.g., the falling rate of profit, the formation of an industrial reserve army, or the tendency toward concentration of capital). Proposition (4) represents Marx's rational-choice analysis of the capitalist's decision-making process in particular circumstances; proposition (5), the move from individual choice to collective behavior; and proposition (6), Marx's effort to discover the institutional consequences of the universalization of the optimal strategy of the typical capitalist. Proposition (6) thus represents the particular institutional logic associated with the specific factors identified in proposition (4).

The "simple reproduction" example considered above might seem to differ significantly from the others. Here Marx appears to engage in macroeconomic analysis without regard to the microlevel mechanisms underlying the macrophenomena. However, this impression is misleading for in his account of the conditions of simple reproduction Marx is merely working out some of the formal conditions entailed by his system of description of capitalist production (the labor theory of value). These conditions must be satisfied in the aggregate if the capitalist economy is to be in equilibrium. However, this discovery by itself has no explanatory significance; it is only when we couple this scheme with an analysis of the mechanisms by which production in the two departments is adjusted so as to satisfy these relations that the account acquires explanatory force. This further examination, however, will involve just the sort of microanalysis characteristic of the first two examples. Such an account might go along these lines: Suppose that conditions of equilibrium are initially satisfied. Then suppose that production of capital goods is increased by some significant amount, with the result that the output of Department I exceeds the demand generated by both sectors for capital goods. The price of capital goods will now begin to fall, since supply is greater than demand. Some capitalists will begin to disinvest in Department I, thereby decreasing the output of Department I. This will lead eventually to a new equilibrium (supposing that new disturbances are not introduced at too rapid a rate for the equilibrating process to catch up). This account explains the fact that capitalist production generally satisfies the equilibrium conditions specified by the schemes of reproduction, but it also depends on just the sort of microanalysis of the conditions of rational choice that was characteristic of the earlier examples.

Here, then, is the model of explanation that appears to be at work in *Capital*. The capitalist economy is defined by a set of social relations of production (property relations). These relations determine relatively clear circumstances of choice for the various representative actors (the capitalist, the worker, the financier). These circumstances are both motivational and conditioning: They establish each party's interests, they determine the means available to each, and the constraints on action that limit choice. The problem confronting Marx is this: to demonstrate, for a given characteristic of the capitalist mode of production (e.g., the falling rate of profit), that this characteristic follows from Marx's account of the defining institutions of capitalism through reasoning about rational behavior within the circumstances of choice they define. Thus Marx's basic pattern of explanation can be summarized in the following way: A given feature of capitalism occurs because capitalists are rational and are subject to a particular set of incentives, prohibitions, and opportunities. When they collectively pursue the optimal individual strategies corresponding to these incentives, prohibitions, and opportunities, the explanans emerges as the expression or consequence of the resulting collective behavior.

Logic of Institutions

This model of explanation is one that has a venerable history. It is similar in form to Hobbes's reasoning about the state of nature.[31] Hobbes makes simple assumptions about rationality and aims and then works out the consequences for a group of a competitive struggle to achieve those aims. We can describe explanations of this sort as being based on a "logic of institutions" that is created by rational individuals pursuing independent ends under structured conditions of choice. A society is made up of a large number of individuals who act out of a variety of motives. These individuals are subject to very specific incentives and conditions that limit their actions and propel them in particular directions. And large-scale social patterns may be explained in terms of the conditions within which individuals make plans and act. Thus social changes and patterns are the result of the actions of many different human agents. A condition imposes an institutional logic on society whenever it provides a constraint or incentive that tends to shape individual conduct so that a pattern emerges.

Lynn White's discussion of the influence of the heavy plow on medieval field shape and property provides a clear example of this sort of "logic." The heavy plow made possible a dramatic increase in productivity and permitted cultivation of the heavier soils of northern Europe, but it required a reorganization of peasant agriculture (the creation of the manor) because it required a substantially greater investment of resources (eight oxen rather than two). Moreover, the size of the plow and its turning radius required the replacement of traditional square fields with a series of long narrow strips. Thus the invention of a more efficient technology imposed a logic of change on the structure of medieval society.[32]

Crucial to this sort of explanation is reference to the incentives and opportunities presented to various participants. If there were no participant or class on the scene with both the power and the incentive to struggle for the social changes necessary to permit the use of the heavy plow, then the plow would not have come into play. Thus it is not sufficient to observe that a given innovation would have beneficial collective consequences to explain or predict its occurrence. It is necessary to identify the particular classes or agents that are so situated as to have both the incentive and the opportunity to introduce the innovation. In fact, instances abound of technical innovations that disappeared because no particular group had both the opportunity and the incentive to introduce them, for example, the waterwheel at the time of the Roman Empire.[33] In this case, the group with the effective power and opportunity – the large slaveholding landlords – lacked a reason to promote the waterwheels, since this class had no incentive to increase labor productivity.

Robert Brenner's analysis of the development of medieval agriculture and changing class patterns represents a paradigm of this sort of analysis. "My argument thus started with the assertion that the feudal social-property system established certain distinctive mechanisms for distributing income and, in particular, certain limited methods for developing production, that led to economic stagnation and involution. It did so, most crudely, because it imposed on the members of the major social classes – feudal lords and possessing peasants – strategies for reproducing themselves that, when applied on an economy-wide basis, were incompatible with the requirements of growth."[34] That is, each major class had a combination of incentives and opportunities that prevented it from effectively pursuing productivity-enhancing innovations – with resulting economic stagnation for society as a whole. Thus agricultural innovations were available – for example, cultivation of larger parcels of land with resulting economies of scale – that did not come into use because none of the relevant classes had both the opportunity and the interest to secure them.

On my interpretation of Marx's explanatory paradigm, the motivational and systemic conditions defined by the social relations of production of capitalism impose an institutional logic on the capitalist economy in this sense: They define the interests that guide various actors within the capitalist economy, and they define the prohibitions and incentives that influence deliberation. They thus represent a highly structured system within which individuals act, and they impose a clear logic of development and organization on the mode of production as a whole. Explanation therefore consists in showing the process through which these conditions stamp the observable features of the capitalist system.

One result of this account of explanation concerns the nature of the conclusions at which this form of reasoning arrives. Marx's arguments give rise to "laws of tendency" rather than "iron laws of development." Once again the disanalogy between astronomy and the theory of capitalism (already discussed in chapter 1) bears emphasizing. Whereas celestial mechanics provides exceptionless "laws of motion" of the planets, Marx's account of capitalism provides only laws of tendency: Other things being equal, we can expect the rate of profit to fall over the medium run. This difference has much to do with the great complexity of factors involved in social change. Marx isolates certain factors from the rest of the social system and deduces their influence on the system as a whole. But because he has abstracted from other causally relevant factors, the behavior of the whole may be expected to differ in some ways from that predicted by the abstractive model. D.-H. Ruben's treatment of the form of necessity contained in Marx's arguments is useful in this context. He discusses the ontological significance of "laws of tendency" and shows that the tendential character of Marx's conclusions is not ontologically significant; it does not correspond to a "tendential" character of social processes. Rather, it is an epistemological fact; it has to do with the limitations of the ability of a social theory to take account of a sufficient number of causally relevant factors to produce nontendential laws. "I think that there is no textual evidence whatever to suggest that Marx thought that tendency talk marked anything other than something about our current inability to state sufficient conditions, and no suggestion whatever that there is no set of conditions sufficient for the occurrence of the phenomenon under investigation."[35]

This treatment of Marx's reasoning gives content to his idea that the capitalist mode of production has a logic of development. This sometimes sounds like a very Hegelian idea (perhaps based on a "dialectic" of development). But the interpretation offered here establishes the consonance of the idea of a "social logic" with other branches of empirical social science by showing how a logic of development can be understood in terms of a logic of institutions. This interpretation therefore offers further support for the view taken in chapter 4 that Marx's system depends neither on Hegel's philosophical system nor on his dialectical method.

Methodological Individualism and Microanalysis

This account makes Marx's explanations rather similar to many other examples from social science — Weber's account of social action, Durkheim's account of mass phenomena such as rates of suicide, or even marginalist economic accounts of price phenomena. Each of these rests on an analysis of the consequences for the social system of the structure of incentives within which participants make choices and act.[36] Does this treatment entail that Marx holds that social phenomena are reducible to individual action? Is Marx a covert methodological individualist?[37]

Marx's stance toward individualism is complex. He is unquestionably opposed to various forms of individualism, in particular, those according to which it is the presocial characteristics of human nature that determine social institutions. Against this form of individualism, Marx holds that individuals can be concretely considered only within the context of the specific social relations that surround them. "Society does not consist of individuals, but expresses the sum of interrelations, the relations within which these individuals stand" (*Grundrisse*, p. 265). That is, there is no such thing as bare individuals; rather, there are only social individuals who find themselves within specific social relations. At the same time, Marx explicitly rejects the idea that society is an entity: "To regard society as one single subject is . . . to look at it wrongly; speculatively" (*Grundrisse*, p. 94). "Let us return to the fiction of the person, Society" (*PP*, p. 96).

Against both these positions—the nonsocial individualism of political economy and the uncritical holism of speculative philosophy—Marx puts forward an alternative position. On this account the socialized individual is the ultimate unit of analysis in social explanation. "Individuals producing in society—hence socially determined production—is, of course, the point of departure" (*Grundrisse*, p. 83). Similarly, in *The German Ideology* Marx writes that "the premises from which we begin are not arbitrary ones. . . . They are the real individuals, their activity and the material conditions under which they live, both those which they find already existing and those produced by their activity" (*GI*, p. 42). And much the same view is advanced in *The Eighteenth Brumaire* as well: "Men make their own history, but not of their own free will; not under circumstances they themselves have chosen but under the given and inherited circumstances with which they are directly confronted" (*SE*, p. 146). On this account "society" is not a freestanding entity, and social relations exist only through the individuals who stand within them. At the same time, however, individuals exist only within particular sets of historically given social relations. Consequently, social explanations must begin with a concrete conception of the individual within specific social relations.

These passages suggest a sensible standpoint on the issue of methodological individualism. For who could deny that social phenomena are inseparable from the actions of the persons who make up society? These agents act under social constraints, however, and the historical specificity of these social constraints allows us to distinguish alternative social patterns of organization. For example, a society in which success and failure are measured in terms of military prowess will have a different dynamic from one in which they are measured in terms of profit and accumulation. Therefore Marx can recognize the centrality of individual action in social explanation while at the same time insist on the irreducibly social character of the conditions that constrain individual action.

What distinguishes Marx's system from non-Marxist social science is not primarily methodological; rather, it is the substantive empirical content of his system. Against "bourgeois" political economy Marx holds that the social relations of capitalism are historically specific and exploitative, and he provides a developed theoretical and descriptive account that attempts to identify those social relations in detail. Thus Marx's distinctive contribution is his account of the specific features of capitalist production that distinguish it from other modes of production and his derivation of the "laws of motion" that follow from these features. But his "derivation" of these laws of motion depends on use of the concept of individual rationality, which he shares with orthodox social science.

The theory of explanation attributed to Marx here offers substantial support for the view that Marxism requires a microanalysis of social processes through consideration of the structured activities of rational individuals. Marx's intention is to explain macrolevel social facts such as the falling rate of profit and the tendency toward technical innovation. But he approaches such problems by considering the factors that constrain and motivate individual activity. Marx's paradigm thus indicates that he accepts the view that the ultimate source of social change is the active individual within specific social relations. Indeed, Marx's view that class conflict is the engine of historical change indirectly expresses this same point; for class conflict proceeds through definite individuals within specific class relations taking action to secure their interests.

This treatment of Marx's view of social explanation finds some support among several recent commentators.[38] Thus D.-H. Ruben discusses Marx's system in terms that are quite compatible with this account of methodological individualism. "I take methodological individualism to be the thesis that statements which appear to refer to any social things can be translated without remainder into statements which actually refer to material individuals and ascribe to those individuals social properties."[39] He argues that Marx's system conforms to this statement of the doctrine of methodological individualism; in particular he denies that the mode of production should be considered an entity.[40] Likewise John Roemer holds that "class analysis must have individualist foundations . . . [and] class analysis requires microfoundations at the level of the individual to explain why and when classes are the relevant unit of analysis."[41]

This discussion has important consequences for the issue of functional explanation in historical materialism. Our account of functional explanation in chapter 2 suggested that Cohen was mistaken in believing that "elaborations" for functional explanations were useful but dispensable, and the main conclusions of this chapter further support that assessment. For our analysis of Marx's practice in *Capital* indicates that reasoning about individual action in constrained circumstances of choice plays a central role in his typical explanations. And this in turn supports the "microanalysis" approach argued by Elster, Roemer, and others against Cohen's functionalist approach. At the same time, the analysis provided here of Marx's explanations suggests that the application of rational-choice theory, game theory, collective choice theory, and so on to the main problems of

Marxist investigation is not a novel development in the hands of Roemer, Cohen, Elster, and others. Rather, it is a more technically sophisticated application of explanatory efforts put forward by Marx himself in *Capital*.

Relation to the Subsumption Theory of Explanation

To what extent does this paradigm correspond to the classical subsumption model of explanation? This model (also described as the hypothetico-deductive [H-D] model) may be formulated as follows: A explains B if and only if A is a set of statements including at least one general law, and A entails B.[42] This account emphasizes that explanations implicitly depend on lawful regularities of various sorts (e.g., laws of nature); to explain an event it is necessary to subsume it under a relevant set of lawlike regularities. On this account, then, explanations presuppose the idea of a law-governed regularity; to explain an event is to show that "given the circumstances A, B necessarily occurred given the relevant laws of nature."

Do Marx's explanations conform to the subsumption theory? That is, may Marx's explanations be represented as deductive arguments whose premises include statements of lawlike regularities and initial conditions? The answer to this question must be qualified. Formally the institutional-logic explanations conform to the subsumption model. They consist of a set of premises that include both lawlike generalizations (e.g., proposition [1] above) and particular statements of fact (propositions [2] and [3]). However, this parallel is ultimately a misleading one (on pragmatic grounds if not on logical grounds).[43] This is so because the deductivist model assimilates the notion of explanation to that of subsumption under a law-governed process of change. On such an account the explanatory force of a given argument depends primarily on the law statements it encompasses. There are statements of lawlike regularities in Marx's explanations, but these statements are somewhat trivial—for example, the statement that capitalists are rational. These law statements are necessary to the argument, but the real weight of the argument lies elsewhere: in the particular details of the circumstances of choice in which capitalists find themselves, and in Marx's reasoning from these circumstances to patterns of collective behavior.

Moreover, these explanations are constructed in a way that is different from that suggested by the H-D model. The process of derivation by which Marx arrives at specific conclusions is nontrivial and involves substantive additional assumptions. The model of Euclidean geometry—so influential within the deductivist theory of explanation—is inappropriate here; we cannot represent Marx's account as a set of hypotheses (the theory) and a set of consequences (the theorems). Rather, the construction of these explanations is simultaneously an application of the theory and its further development. Thus the deductivist model oversimplifies the relation between the fundamental analysis of the social system and the explanatory consequences to which this analysis gives rise. It suggests that the relation between explanandum and explanans is a simple deductive one, where the premise entails the conclusion trivially. On the account under consideration here there is an inferential relation between explanandum and explanans, and it can be reconstructed as a simple deductive one, but this misrepresents the process of discovery. The process of discovering an institutional logic is not merely one of working out the deductive consequences of the theory; it is rather discovering new aspects of the social process. These aspects are perhaps "implicit" in the original theory, but their discovery is a substantive one, not a mere deductive exercise.

6

Evidence and Justification

This chapter and the next consider the empirical justification for Marx's analysis of capitalism. Theories of justification in the philosophy of science display a wide range of sophistication and refinement;[1] nonetheless a central part of any acceptable theory of justification is an account of the relation of empirical evidence to the theory, and of the process of inference by which *that* evidence supports *this* theory. Whatever other features a well-justified theory must possess, it must be supported by an appropriate array of empirical evidence. We will take this as a minimal condition on any epistemology of science and will ask what the empirical basis is of Marx's analysis of capitalism. We will consider the chief model of justification offered by philosophers in describing the natural sciences (the hypothetico-deductive model) and will conclude that this model does not apply very successfully to Marx's practice.[2]

Empirical Evidence and Justification

How might we evaluate the truth of a scientific proposition? The basic requirement of justification is that scientific beliefs must be grounded in a body of relevant empirical evidence.[3] Many propositions may be tested directly through observation or experiment. Consider these examples: The earth is 24,830 miles in circumference. Uranium is radioactive. The population of New York City is approximately 8,000,000. Light travels 186,281 miles per second in a vacuum. Each of these sentences expresses a factual assertion for which experimental and observational methods of verification exist. The early logical positivists made a great deal of this simple fact; they sought a set of sentences (observation sentences) whose verification was immediate and infallible that could function as the empirical foundation of all scientific beliefs. Philosophers of science have since recognized that no such sentences exist. In particular, each of the preceding examples can only be verified using procedures that themselves presuppose assumptions we can*not* experimentally verify. To measure the speed of light, for example, we must use instruments in whose observations we can trust only by presupposing the truth of other areas of scientific theory. This qualification establishes that there is no such thing as a *pure* observation sentence. Nonetheless many scientific beliefs have observational standing in this qualified sense: *If* we take as given large parts of background theory, *then* these sentences may be confirmed or falsified through observation and experiment.

The crucial fact is that many scientific beliefs are not observational in even this qualified sense. Consider these examples: Electrons have spin. The human visual system embodies a digital-image processor. Organic evolution is punctuated by periods of rapid species change. These propositions are theoretical; that is, they represent hypotheses that have implications for observation but that cannot be evaluated directly through observation. Supposing that such propositions are important elements of scientific knowledge, how may these be given empirical justification?

The most common answer philosophers of science have given to this question has taken the form of the hypothetico-deductive (H-D) model of justification. A theory (in conjunction with other parts of scientific knowledge) entails infinitely many consequences, some of which are experimentally testable. Scientists use their ingenuity to derive some consequences that are testable; they then perform the experiments. If the experiments turn out as theory predicts, the theory is corroborated. And if the theory has been rigorously tested in this way in a variety of different areas, the theory is justified. That is, the justification of a scientific theory is a matter of working out its observational consequences and testing them directly; the truth of these consequences lends indirect (inductive) support to the theory that gives rise to them.

In addition to this requirement of predictive success, philosophers of science generally identify several secondary criteria of theory justification. These include simplicity, explanatory scope, generality, support from other areas of science, and ontological parsimony.[4] Such criteria are secondary in the sense that they are only relevant in choosing between two theories having approximately the same empirical strength. It would be manifestly irrational to choose a simple, powerful, and parsimonious theory that entailed a great many false conclusions over a complex, weak, and ontologically extravagant theory that entailed only true conclusions.

This account of justification attaches a good deal of weight to the category of observational evidence. Observation functions as the source of empirical content and justification for scientific theories. Observation sentences may be tested fairly directly, and they in turn serve to test the theories from which they are derivable. The idea is that we get a firm grip on the world through observation sentences and then try to establish the truth of our theories through their connections to these observations. Positivists took this position to its logical extreme.

For them observation is not only less problematic than theory; it is epistemically privileged because it is amenable to immediate verification. It might seem, however, that this reliance on observation in the justification of scientific belief has been discredited by antipositivist philosophers of science. The antipositivist turn in the philosophy of science has directed much of its fire against this conception of observation. Philosophers including Norwood Hanson, Thomas Kuhn, and Paul Feyerabend have drawn very radical conclusions concerning observation: All observation is theoryladen, and theories consequently generate their own supporting evidence.[5] Given these criticisms, we must consider whether an epistemology based on empirical evaluation is still credible.

These arguments against the fixity of evidence are central for contemporary philosophy of science. I will take the view, however, that the antipositivist position represents an overcorrection of the admitted excesses of logical positivism. These philosophers have correctly shown the inadequacies of the positivist conception of evidence, and they have permanently overthrown the idea of a strict distinction between observation and theory. However, their own position leads to the undermining of the rationality of science, since it makes it impossible to provide empirical evaluation of scientific belief. It must therefore be qualified by renewed commitment to standards of empirical evaluation.

I will consequently presuppose a more moderate posture based on these assumptions. First, there is no such thing as theory-independent observation and therefore no such thing as a purely empirical foundation for scientific knowledge. Secondly, however, there is a real distinction between observation and theory in terms of the applicability of observational and experimental procedures in testing the proposition. Sentences that are relatively observational on this criterion constitute the basis of justification of scientific belief. This position represents a rejection of much of the positivist program, since it rejects the aspiration to a theory-independent empirical foundation for scientific knowledge. What remains is a more reasonable epistemological stance that seeks to identify a body of belief that is provisionally fixed relative to current theory, with the result that the theory can be assessed in terms of it. To justify one part of science we almost always make use of other areas of scientific knowledge, so there can be no absolute justification of theory. At the same time, however, empirical evidence retains primacy in assessing scientific belief.[6]

Marx's Logic of Justification

Let us turn now from these general considerations to an examination of Marx's justificatory practice. What sort of justification does he offer in support of his analysis of capitalism? It was argued in chapter 1 that Marx's construction is not a unified theory in which all aspects relate back to the central hypotheses. Rather, it includes a variety of types of analysis and description:

1. An abstract description of the defining structural and functional properties of the capitalist economic structure.
2. An analysis of the institutional logic created by these features of the social system.
3. A developed theory of value that functions as a tool of analysis.
4. Considerable economic analysis.
5. A collection of sociohistorical passages: "The Secret of Primitive Accumulation" and "The Working Day," to name two.

How may such an account be empirically evaluated? What counts as evidence? Is a deductivist model appropriate? In order to answer these questions it is necessary to consider Marx's actual epistemological practices in some detail: the ways in which he makes use of empirical data, the sorts of arguments he puts forward in support of his main hypotheses.

In general I will maintain that Marx makes careful and detailed use of empirical data in evaluating his construction, but I will also hold that evidence is not relevant to his analysis in the way the H-D model requires, that is, as a body of evidence to be subsumed under a unified deductive theory. Rather, Marx's practice of justification contains at least the following parts. First, a good deal of his account is subject to independent empirical evaluation—that capitalism is a class system, for example. In particular, we can empirically and historically evaluate Marx's description of the capitalist mode of production (CMP) and its defining structural and functional properties. Property, profits, accumulation—these features of the capitalist system may be investigated through both ordinary and specialized knowledge about capitalism. Is capitalism a class system? Is Marx's account of the structure of capitalist property substantially correct? Is accumulation the rule in capitalism? These questions can be considered on the basis of concrete empirical and historical investigation. This is not to say that Marx's position on these questions is beyond controversy. Historians and social scientists may differ in their reading of the evidence. But these disagreements will turn on fairly concrete facts about the social system. This aspect of Marx's justificatory practice conforms to the point argued in chapter 1 that his analysis is largely descriptive rather than theoretical or hypothetical.

The second half of the justification of Marx's account has to do with the rigor of his reasoning. Much of *Capital* consists in working out the logic of the institutions uncovered in his basic account of the CMP. This process of reasoning can be evaluated on internal logical grounds: Are Marx's arguments rigorous? Does he take account of the relevant factors? Do the simplifying assumptions he introduces crucially affect the outcome? Are the inferences he makes sufficiently precise and careful? Marx is attempting to determine the implications for the institution of a set of identifiable background conditions. If his reasoning is weak, then we will have little confidence in his conclusions. This is not a form of empirical evaluation at all; it is conceptual and logical, and makes it possible to assess the degree of credence to attach to Marx's conclusions.

Finally, Marx's analysis gives rise to predictions about the behavior of capitalism. This suggests that it should be possible to evaluate the analysis on deductivist grounds: Are its predictions borne out? I will argue, however, that this form of justification is of the least usefulness for evaluating Marx's system because of general difficulties in making definite predictions in social science.

On this account, then, Marx's analysis of capitalism acquires epistemic warrant through (1) the independent confirmation available for his basic premises, and (2) the rigor of his reasoning from these premises to the institutional logic to which they give rise. Only derivatively is it necessary to consider the long-term consequences of the analysis; the truth or falsity of Marx's predictions provides only the weakest possible test of the correctness of his treatment of capitalism. Consequently the H-D model does not suit Marx's practice, and charges against him that presuppose that model miss the mark widely. Marx is offering not a general theory but an explanation of capitalism based on a variety of independent premises.

Marx's Epistemological Practice

Let us first consider a few examples of Marx's use of empirical data. *Capital* begins with a discussion of the commodity form, so let us begin by considering the empirical basis of Marx's description of capitalism as a system of commodity production. In the opening chapters of *Capital* Marx does at least two different sorts of things. First, he identifies some of the central features of an exchange economy: For example, goods are produced for exchange rather than for immediate consumption (*Capital I*, p. 131); goods exchange at known ratios (pp. 139–40); commodities have both use value and exchange value (p. 126); systems of exchange generally depend on a medium of exchange (money) rather than simple barter (pp. 188–89); commodity production represents one particular form of the social division of labor (p. 133); commodities are produced through the expenditure of human labor power (p. 128); and the price of a good falls when labor productivity rises (p. 202). These are the defining characteristics of a commodity system (as opposed to an economy based on immediate consumption or fully socialized production). Furthermore these aspects are advanced as expressing common knowledge about capitalism. They need no refined empirical justification; on the contrary, any participant in an exchange economy will recognize the correctness of these claims.

The second aspect of Marx's work in these chapters is formal and analytic.

He introduces the distinctions in terms of which he will characterize the capitalist process of production throughout the book, and he reasons from the simple empirical qualities of the commodity form to their institutional presuppositions. Thus he distinguishes between abstract and concrete labor (p. 132); he introduces the concept of value as embodied labor time (p. 128); he presents his concept of simple average labor (p. 135); and so forth. These stipulations are not empirically grounded; they are part of the apparatus Marx uses to analyze the process of production as a whole. The epistemic practice in this chapter, then, is quite simple. On the one hand Marx refers to empirical features of capitalist production that represent ordinary firsthand acquaintance with capitalism. On the other hand he makes a series of analytic points that will remain formal and nonempirical until they are developed in later chapters into a fullscale examination of capitalist production.

Consider next Marx's substantial use of empirical matter in his description of the effects of industrialization on the working class. Here Marx relies on the *Reports of the Inspectors of Factories* and other descriptive sources to document contemporary conditions. He recounts the extensive use of child labor (*Capital I*, p. 592), the appalling health conditions found in the lace industry (*Capital I*, pp. 595–98), the overwork of industrial workers not covered by the Factory Acts (*Capital I*, pp. 608–10), and so forth. In these passages (and parallel text in chapter 10 on the workday and in *Capital III*, pp. 182–90), we find that Marx presents a great deal of factual and statistical data to document the conditions of life and work associated with industrial capitalism. This information offers concrete confirmation for some of the explanatory points Marx argues in these chapters: that capital tends to intensify and extend the conditions of labor; that it tends to introduce technologies that increase the productivity of labor at the expense of workers; and that it tends to replace adult skilled labor with unskilled and child labor.

A third important example of Marx's epistemological practices occurs in his treatment of various of the "laws of accumulation" of capitalism: the tendencies toward concentration and centralization, increasing productivity and technological innovation, and the formation of the industrial reserve army (*Capital I*, chapter 25). In this chapter Marx continues to use the ordinary participant's knowledge of capitalism, but he supplements this with more specialized economic and statistical data. For example, in his treatment of the formation of the industrial reserve army, Marx uses census data from England and Wales to confirm the shift in employment patterns predicted by his argument; these figures show that most industrial occupations decreased in employment between 1851 and 1861, whereas employment in mining and cotton work increased (*Capital I*, p. 783). Marx uses these data to support his contention that technological innovation periodically displaces large numbers of workers into the industrial reserve army. Likewise, Marx puts forward relevant statistical information to assess the behavior of the wage during the period of 1849–59 (*Capital I*, p. 791), demonstrating that the industrial reserve army has the effect of depressing the wage.

A related example occurs in Marx's discussion of the effect of industrialization on levels of employment in the capitalist economy (*Capital I*, pp. 575–88). Here his use of data is sharply focused on an important theoretical point: What are the probable consequences of industrialization on levels of employment? Marx's own analysis suggests that each round of technological innovation will cast large numbers of workers into the ranks of the industrial reserve army, whereas the view of orthodox political economy is the reverse. In this passage Marx considers a wide range of information on the actual movement of levels of employment in various British industries, and the specific character of technical innovation in the context of rapid accumulation of capital. He surveys cycles of depression and stagnation from 1770 to the mid-nineteenth century, showing how this history corresponds in detail to his own account of the process of innovation and accumulation (*Capital I*, p. 583). This passage represents a paradigmatic example of Marx's ability to combine his treatment of the capitalist system of production with a volume of specific, relevant empirical data.

Finally, consider Marx's use of data from the history of capitalism throughout *Capital*, and particularly in part eight ("So-Called Primitive Accumulation"). Here Marx provides an account of the early development of the social relations of production of capitalism and the emergence of these relations from the postfeudal period. In these chapters he presents an analytic hypothesis, namely, that capitalism is a social order that depends on quite specific social preconditions. "One thing, however, is clear: nature does not produce on the one hand owners of money or commodities, and on the other hand men possessing nothing but their own labour-power. This relation has no basis in natural history, nor does it have a social basis common to all periods of human history. It is clearly the result of a past historical development, the product of many economic revolutions, of the extinction of a whole series of older formations of social production" (*Capital I*, p. 273). Central among these historically specific preconditions is the separation of the immediate producer from the instruments of production and the accumulation of capital in private hands. "Capitalist production therefore reproduces in the course of its own process the separation between labour-power and the conditions of labour. It thereby reproduces and perpetuates the conditions under which the worker is exploited" (*Capital I*, p. 723). In the chapters on primitive accumulation Marx provides this general view of the historical specificity of capitalism with concrete historical content by outlining some of the processes by which these social preconditions became established in England. He considers a variety of data in this context: a survey of English legislation against vagabonds, paupers, and the rural poor during enclosure (*Capital I*, pp. 896–904); accounts of the process of dispossessing the English freeholding peasantry (*Capital I*, pp. 877-95); data concerning the early formation of industrial capital (*Capital I*, pp. 908-26); and so forth.

Varieties of Evidence

These examples give some idea of the variety and scope of empirical data to which Marx appeals. In this section we will examine more closely the fund of empirical evidence on which Marx's analysis of capitalism is grounded. What facts about the capitalist system does he regard as providing the basis of empirical evaluation for his analysis?

First, Marx relies on a wide variety of knowledge about current features of capitalism that must count as observational. As we saw in our discussion of the empirical basis of Marx's account of the commodity form, he uses a great deal of data about capitalism that is grounded in ordinary firsthand acquaintance. Any competent observer or participant in a capitalist economy will recognize that capitalism is based on economic rationality, production for profit, competitive markets, wage labor, and a division of classes over the ownership of private property. We know that the system must reproduce itself over time; that it is a system of production for exchange; that it is a wage system; that commodities exchange at publicly known and relatively stable ratios; that money is used to represent these exchange ratios; that interest is paid on borrowed money at a publicly known rate; that capital receives a publicly known rate of profit; that rent, profit, and interest are related insofar as all represent the return on the investment of wealth; that wages, prices, and employment levels vary over time; and so on almost indefinitely. This sort of evidence is used throughout *Capital*, for example, in Marx's discussion of the commodity form (considered earlier), in his description of cooperation and industry (*Capital I*, pp. 444-49), and in his account of the character of the division of labor (pp. 458-61).

Separable types of claims are being made here, but their common epistemological status is that they all reflect beliefs any competent observer would accept through an ordinary acquaintance with capitalism. Such information can therefore be classified as observational. These beliefs are incontrovertible, but they can be asserted with reasonable confidence. They thus function as a form of observational evidence: knowledge we possess about the CMP that is relatively unproblematic.

Marx supplements this form of knowledge with more specialized forms. Extensive sources in political economy and social description provide him with considerable factual information about the current structure of the CMP. A great volume of specific data on various economic categories is available from these sources: data concerning the nature and behavior of ground rent (*Capital III*, chapters 38-45); evidence on the forms taken by the wage within industrial capitalism (*Capital I*, pp. 683-706); data detailing the role of finance within a mature capitalist economy (*Capital III*, chapters 16–19 and 21); and so forth. Thus Marx derives from the classical political economists an analysis of the underlying structure of the CMP—its reliance on wage labor, its economic rationality and orientation to profit, its industrial organization, its basic process of clarification of classes into workers and owners. The basic structure of the CMP can be pieced together from these sources: production for profit, wage labor, increasing centralization and concentration, the evolution of systems of money and finance, and the distribution of profits among industrial, agricultural, and finance capitalists. The phenomena of capitalism are presented in great detail in the literature of classical political economy; Marx synthesizes these various sources and constructs a single analysis of the underlying system of production.

Marx also makes extensive use of social descriptive writings. These are important for filling in the contemporary condition of the social system under capitalism. They allow Marx to complement his historical perspective on the development of capitalism with a detailed view of its present condition. Thus the statistical and descriptive evidence concerning the condition of the lower classes in nineteenth-century England supports Marx's theoretical account of mature capitalism. It sheds harsh light on the historical process of class formation within capitalism and on the striking contradictions of capitalism: its creation of tremendous social wealth and individual poverty. Thus Marx uses contemporary sociological writings to support the class analysis he incorporates into the political economic description of capitalism. Having shown that profit is impossible in the absence of wage labor, he marshalls contemporary descriptions of English social stratification to support the depiction of capitalism as a system that rests on a division of society into two broad classes: owners and nonowners.

A final large category of information to which Marx appeals is historical. Certainly this sort of knowledge is far from self-evident. It is theory laden in that even the broad outline of historical development may be reasonably disputed. Marx presupposed one plausible interpretation of this segment of European history and offered it as substantiation for his analysis of capitalism. It is possible to provide support for this view of history, but evidence of this sort is plainly not observational in the narrow sense. This treatment of historical data as observational is quite defensible, however, in light of contemporary skepticism concerning the possibility of any range of evidence that is wholly nontheoretical. We cannot begin afresh in every science; we must accept—at least provisionally—the results of other areas of scientific knowledge. This is true of every science, not merely Marx's. So Marx is justified in treating other forms of knowledge as support for this theory, even though they are not epistemologically primitive. Such evidence serves to establish much of the detail of the features of capitalism as an economic system since its emergence.

Let us consider Marx's use of historical data more closely. First, he has detailed information on the chronology, social structure, legal system, and political development of the Roman Republic and Empire. This gives the category of the mode of production a good deal of its historical content and adequacy, for the evidence available to Marx about classical Rome gives enough detail about its economic and social organization to formulate a fairly comprehensive description of its system of production. This means that the concept of the mode of production is historically informed and not the product of purely theoretical or philosophical analysis (analysis Marx labels historico-philosophical; *Correspondence*, p. 294). The Roman system constitutes a distinctive organization of the productive resources of society–a system based on slave labor, a class division between patrician, plebeian, freedman, and slave, a system based on commerce and urban production without capitalism and wage labor–in short, a specific and detailed alternative to the capitalist mode of production. These materials provide an opportunity to analyze a concretely described historical social formation, and they provide much of the empirical and historical content of the central ideas of historical materialism. They permit Marx to develop both the framework of a class analysis and the distinction between the economic structure and the institutions of law, state, and religion. This evidence affords the historical content and the testing ground for the important concepts of the mode of production, the account of classes, the primacy of the institutions of production in social explanation and historical development, and so forth.

Second, Marx has informed knowledge of the social structure of medieval Europe and the feudal system of production and social organization. This evidence is important to his analysis of capitalism in two ways. First, it provides yet another historically determinate example of a mode of production, structurally distinct both from the Roman and the capitalist example. So it gives further content to the concepts of the mode of production and the economic structure. Second, the feudal information is particularly important because feudalism is the historical origin of capitalism: Capitalism developed in the context of feudal social relations and ultimately transformed those relations. For this reason a historically detailed analysis of the social relations of feudalism and their dynamic for change is of great import for the question of the emergence and development of capitalism.

The German Ideology demonstrates this quite concretely; there Marx traces in detail the evolution of commercial capital in feudal towns and the dynamic that development imposes on the growth of the system of property ownership through medieval guilds, merchant capital, manufacturing capital, and finally fully developed bourgeois private property and industrial capital (*GI*, pp. 68–79). This analysis is conducted in fairly specific historical detail and provides a foundation for Marx's theoretical analysis in *Capital*. An account of feudalism and medieval social structure thus affords historical and empirical evidence of significance to the mature theory of capitalism, because it offers a detailed view of an alternative form of social organization and of a form of social organization that develops into capitalism. This latter fact allows Marx to shape the basic categories of his analysis of capitalism historically, showing the emergence of the basic characteristics of capitalism. Thus he traces the development of systems of forced labor from feudal bondage to yeomen's independence to wage labor; of capital from feudal wealth to commercial capital to industrial capital; of the institutions of the state from absolute monarchy to the liberal state; and so forth (*GI*, pp. 43–46, 68–86).

Finally, Marx uses substantial information concerning the historical development of modern capitalism: its emergence from feudalism, its tendencies toward mechanization and concentration, its tendency to expansion, its consequences for the working class, and the like. Here evidence is available concerning both the historical development of capitalism and its modern characteristics of organization. Marx pays special attention to the transformation of agriculture from small holdings to more rational large-scale holdings and its consequent effect of dispossessing the peasantry (*Capital I*, pp. 873–940). The Enclosure Acts represent the political aspect of this process. He also notes the development of capitalist production through its successive stages: handicraft industry transformed into manufacture, manufacture transformed into factory, and factory evolving into a mature industrial system. The evidence drawn from economic history is of special importance here, particularly for the light it sheds on social organization and the transformation of social relations through the development of capitalist production. Here the history of prices and wages is of some importance, for it allows an informed estimate of the conditions of life in various stages of the development of capitalism.

Data drawn from the history of the emergence of capitalism supports Marx's analysis of the fundamental relations of production of the CMP. Marx is able to use evidence from contemporary English history to show how the class formations of medieval England developed into the modern class distinction between capital and labor. Thus the independent peasantry was increasingly dispossessed of its traditional right to the use of land and was therefore transformed into the urban poor, while the landed aristocracy was forced through the decline of rents into an accommodation of and eventual incorporation into capital, through the capitalization of land. History therefore displays a pattern of the formation and clarification of two great classes, the proletariat and the capitalist class, which supports Marx's abstract description and class analysis.

These empirical and historical sources of evidence give Marx a basis for evaluating his description of the defining features of the capitalist mode of production: the structural and functional characteristics that distinguish it from other systems of production. Support for this description is drawn from all three sources: history, sociology, and political economy. Marx's account of capitalism as a pure mode of production is founded on his detailed knowledge of its actual historical development, beginning with the institutions of commerce originating in medieval towns. He traces the growth of commodity economies, shows how they differ from the Roman example, and demonstrates the dynamic by which they develop into a fully articulated system of wage labor. Thus Marx uses the history of medieval Europe to support his basic analysis of capitalism as a system of production by showing how it evolved coherently from earlier forms of organization. Second, he uses the work of the most advanced political economy to extract the basic characteristics of capitalism as a contemporary system of organization: the market system, profit rationality, competition, the distinction between use of value and exchange value, the theory of money and credit, and so forth. Political economy provides Marx with a sophisticated description of the surface structure of capitalism and a wealth of empirical evidence supporting that description.

Marx's abstract examination of the structural and functional properties of the capitalist mode of production thus rests on a massive integration of evidence drawn from European history, sociological studies of contemporary social conditions, and sophisticated descriptions of its present organization drawn from political economy. Such information can properly be regarded as the fund of evidence in terms of which to evaluate theories of the CMP. Moreover the premises of Marx's abstract analysis can be evaluated largely independently. Thus the justification of these premises is substantially nontheoretical in a crucial sense: Their empirical standing does not rest on the further articulation of the theory, but rather on empirical and historical arguments independent of the rest of Marx's account of capitalism. Marx has seriously considered the appropriate range of empirical and historical information, and this evidence offers a reasonable basis for empirically evaluating his analysis. Thus we can conclude that Marx possesses a store of observational evidence that can be used to assess his basic premises.

Rigor of Analysis

Let us turn now to a second aspect of Marx's justificatory practice. This concerns his effort to work out the institutional logic of the defining institutions of capitalism. It is a form of justification and evaluation prominent in social and economic analysis but not in natural science. This criterion of evaluation pertains to the rigor of the analysis. Are the author's arguments convincing? Are they adequately detailed? Do they overlook factors that ought to be taken into account? This is a form of internal evaluation. Some economists are superior to others in the rigor and care with which they make their cases. Here several standards are applicable. First, are the economist's inferences sufficiently precise and careful? This form of evaluation is similar to checking a proof in mathematics; it treats the analysis as something like a derivation and assesses the reasoning. Second, are the premises of the analysis – the hypotheses with which the economist begins – reasonable and empirically defensible? And third, is the overall thrust of the analysis consistent with the features of the economy that we already "know"?

This aspect of Marx's epistemological practice bears a close resemblance to J. S. Mill's "deductive method" in *A System of Logic*.[7] Mill presents the deductive method as an alternative to inductive or hypothetical methods of confirmation. It is one according to which a science determines the social tendencies of a set of abstract laws, without being able to verify the laws through their predictions. Rather, the laws themselves are justified more or less directly. Mill believes that the deductive method is the only one possible in inexact sciences like economics where predictions are almost always heavily qualified by background assumptions that may only seldom be completely satisfied.[8]

This form of justification corresponds to the view taken in chapter 5 that Marx's analysis is chiefly concerned with working out an institutional logic. He is attempting to arrive at the implications for the institution of a set of identifiable background conditions. If his reasoning is weak, we will have little confidence in his conclusions. Consider an example that illustrates each aspect of justification on this account. Marx argues that capitalism necessarily tends to expand its scope into new regions of the world, because the drive for accumulation impels the capitalist to seek external markets. This stance may be evaluated on all these grounds. Does Marx make the case in a logically compelling way, given his starting point? Is the assumption that capitalists are driven to accumulate a defensible one? And do capitalist economies generally tend toward one form of expansion or another?

Note that this form of justification is quite different from the standard view of justification of scientific theory (the H-D model). It is a pattern of inference that transmits empirical weight to conclusions through fairly esoteric chains of reasoning rather than through any direct or indirect test of the conclusions themselves. In practice this is a common form of evaluation in economics generally (and perhaps other social sciences as well).

Marx's economic analysis has received mixed reviews from economists. Some modern economists like Paul Samuelson have a somewhat disdainful attitude toward Marx's economic work,[9] but Samuelson's criticisms of Marx are based on a highly simplistic understanding of Marx's system.[10] Wassily Leontief takes a more measured attitude; he acknowledges Marx's significant achievements in his description of capitalism but attributes these to Marx's "realistic, empirical knowledge of the capitalist system" rather than to his analytic accomplishments.[11] More significant is Joseph Schumpeter's appraisal, however, since he is a respected orthodox economist who knows Marx's work well. Schumpeter's judgment is that Marx is a gifted economist who develops the Ricardian system rigorously and fully, and he cites a number of achievements as having enduring importance.[12] He regards Marx's analytic skills as significant and exacting. Finally, Michio Morishima regards Marx's analytic achievements as massive and important.[13]

Deductive Consequences of the Analysis

The hypothetico-deductive model of justification (the H-D model) maintains that scientific theories are evaluated through experimental evaluation of their observational consequences. Is this an accurate description of Marx's practice? In this section I will argue that it is not. Marx does indeed draw deductive consequences from his analysis—such as the "laws of motion" of the capitalist mode of production; these take the form of predictions about the behavior of the capitalist economy. These consequences are often seen as critical to the empirical standing of Marx's construction. Defenders of Marx like Ernest Mandel regard evidence of this sort as straightforward confirmation of the truth of Marx's analysis: "It is precisely because of Marx's capacity to discover the long-term laws of motion of the capitalist mode of production in its essence . . . that his long-term predictions—the laws of accumulation of capital, stepped-up technological progress, accelerated increase in the productivity and intensity of labour, . . . —have been so strikingly confirmed by history."[14] Critics like Karl Popper, by contrast, regard the failure of various predictions as falsifying Marx's account.[15] However, both positions are mistaken. These predictions do not function as a chief source of empirical confirmation for his account.

It should be noted that by this approach I do not intend to disassociate Marx's construction from its long-term predictions. Marx certainly believes that capitalism will show a definite pattern of development and that *Capital* lays bare the logic that gives rise to that pattern. Thus in the preface to *Capital* Marx writes, "It is not a question of the higher or lower degree of development of the social antagonisms that spring from the natural laws of capitalist production. It is a question of these laws themselves, of these tendencies winning their way through and working themselves out with iron necessity. The country that is more developed industrially only shows, to the less developed, the image of its own future" (*Capital I*, pp. 90–91). And in the penultimate chapter of volume I Marx writes of the necessary collapse of capitalism, "This expropriation [of capitalists by capitalists] is accomplished through the action of the immanent laws of capitalist production. . . . The centralization of the means of production and the socialization of labour reach a point at which they become incompatible with their capitalist integument. This integument is burst asunder. The knell of capitalist private property sounds. The expropriators are expropriated" (*Capital I*, p. 929). Between these boundary points we find a host of smaller-scale predictions concerning the occurrence of crisis, the behavior of the rate of profit, the formation of classes, the restructuring of capital over time, and the like. Indeed, much of *Capital* represents Marx's efforts to develop the main consequences of his account of the capitalist mode of production (outlined in chapter 1): the structural and dynamic effects of the institutional logic created by the essential features of the CMP.

Thus Marx does in fact derive predictions from his analysis as a whole – the "laws of motion" of the CMP, and these consequences represent much that is of interest in *Capital*. They include the tendency of the economy toward concentration, centralization, rationalization of the process of production, extension of the capitalist market into new regions of the world, the endemic crises associated with unbalanced and unplanned production and competition, and the declining rate of profit. Moreover it is plain that Marx believes that his account gives us strong reason to expect that these predictions will be borne out.

What I will argue, however, is that the epistemic role of prediction in Marx's system is the reverse of its role in the natural sciences. Rather than predictions being the basis of justification of the full analysis of capitalism, Marx's forecasts in *Capital* are themselves justified by the independent warrant available for the basic account. Therefore the success or failure of predictions is only the weakest form of test for the basic analysis. This suggests that Popper's charge of unfalsifiability misses Marx in the most fundamental way: It misconstrues the epistemic role of long-term predictions in Marx's analysis.

Earlier chapters have already provided a basis for describing the predictions to which Marx's study gives rise. In chapters 1 and 3 we surveyed some of the main elements of Marx's examination of capitalism. There it was found that his account involves the description of a variety of the institutions and social relations of production underlying the capitalist economy. This portrayal constitutes the foundation of the predictions Marx makes in the context of his scientific study of capitalism. In chapter 5 we considered the logical form of the explanations Marx constructs on the basis of this analysis and, consequently, the logical basis of his forecasts. There it was found that Marx's explanations are rigorous deductive arguments through which he attempts to derive various collective properties of capitalist development from his description of the defining relations of production. These arguments attempt to identify the logic of institutions created by capitalist relations of production. Central among the predictions to which this sort of analysis gives rise are what Marx refers to as the laws of motion of the CMP: tendencies toward a falling rate of profit, technological innovation, centralization and concentration of capital, and economic crisis; the creation of an industrial reserve army; and so forth. (See chapter 5 for specific discussion of these laws.)

These points make it plain that Marx's construction is logically committed to a family of predictions. His analysis of capitalism provides the basis for such predictions, and Marx invests a good deal of energy and rigor in developing the consequences of his account for the long-term behavior of the capitalist system. These predictions for the long-term development of capitalism include both the laws of motion Marx himself emphasizes and tendencies that others have derived using his model. On the H-D model, consequences like these are the heart of the empirical evaluation of a theory. Can we conclude, then, that Marx's system is to be evaluated in terms of the truth or falsity of these predictions?

This conclusion is unwarranted on several grounds. First, Marx does not behave as though his theory were justified on these grounds. If the empirical basis of a scientific theory is the predictions it generates, we would expect scientists to make extensive efforts to empirically evaluate the predictions in detail. However, Marx generally seems unconcerned about establishing the truth of his predictions as a test of his analysis. He does not canvass the relevant economic data to assess the degree of fit between important forecasts and actual economic circumstances. For example, in volume III of *Capital* Marx derives the prediction that the rate of profit tends to fall in periods of capitalist development and technological innovation. He does not, however, consider comprehensive data concerning the movement of the rate of profit in order to verify this prediction in detail (*Capital III*, chapters 13 and 14). These sorts of data were available; it would have been possible for Marx to identify several periods of rapid capital development and create a time-series analysis of the behavior of the rate of profit in these periods. Such an investigation would establish whether there is an empirical correlation between widespread capital development and a drop in the rate of profit. However, Marx is content instead to refer to general, unspecific information about the behavior of the rate of profit. Thus he describes the relevant empirical phenomena in a single sentence: "The rate of profit has a tendency to fall" (*TSV II*, p. 438). He regards this law as a simple empirical fact wellknown to participants and observers in the capitalist system, and one that stands in no need of elaborate empirical confirmation. Instead, the critical problem is to provide an explanation of the fact: "These economists perceived the phenomenon, but tortured themselves with their contradictory attempts to explain it" (*Capital III*, p. 319). Likewise, Marx derives large-scale predictions concerning the formation of the industrial reserve army, but he does not then attempt to verify these predictions in detail by considering a wide range of relevant employment data, in correlation with the sorts of factors he identifies as stimulating growth in the pool of the unemployed (*Capital I*, pp. 780–802). This lack of concern with specific empirical data in connection with predicted regularities makes advocates of the deductivist model of justification uneasy. (These points lie at the heart of Popper's charge of unfalsifiability against Marx.) It makes Marx seem unconcerned with empirical evidence generally.

Second, Marx's "predictions" are deliberately loose in a way that makes them unsuited for justification: Marx also specifies reasons for their failing to emerge, in the form of countervailing tendencies and competing causal factors. As was shown in chapter 1, Marx speaks of "laws of tendency" rather than strict regularities, for example, the tendency for the rate of profit to fall and the tendency for capitalist production to give rise to a pool of unemployed workers. Thus in analyzing the law of the falling rate of profit Marx identifies a series of counteracting factors that work to offset the falling tendency in the rate of profit: more intense exploitation of labor, reduction of wages below their value, drop in the value of capital goods, extension of foreign trade, and so on (*Capital III*, pp. 339–48). These factors all tend to elevate the rate of profit; so the net result is that Marx asserts only that the rate of profit will show a *tendency* to fall, but a tendency that may be counteracted by competing factors. A tendential prediction, however, is not refuted by the nonoccurrence of the event in question, for tendencies may be offset by competing causal factors. Thus Marx's position is consistent with both the occurrence and the nonoccurrence of the predicted outcome. In light of this looseness, it is a serious question whether these predictions have justificatory weight at all.

Finally, a crucial difference between social and natural science makes the deductivist model implausible on its face as a technique of empirical evaluation in social science. This follows from a consideration Popper himself emphasizes in his critique of historicism: the complexity and indeterminacy of social phenomena. It is unreasonable to expect precise long-term predictions from social science. Social phenomena are too complex, with too many interrelated forms of causation and agency at work, to permit confident prediction. The metaphor of the motions of the stars – attractive as it is to Marx – is not useful in social science, since the determinacy of mechanics is precisely what is impossible in social science. Rather, predictions take a conditional form: *If* other factors do not intrude, *then* these tendencies will emerge. But of course other factors generally will intrude, so H-D testing is not of much use in social science. This general feature of social explanation entails that the logic of justification characteristic of the natural sciences is less wellsuited to the social sciences; the falsity of an empirical prediction cannot be translated into the implication that the entailing theory is false as well.

Thus Marx's lack of concern for verifying his predictions in detail becomes understandable if we reject the deductivist model and argue for an alternative form of justification. Given that Marx's predictions take the form of laws of tendency, we can account for their failure on the basis of factors that are not represented in his analysis. And if Marx does not assert that his account is a complete closed system that identifies all causally relevant factors, predictive failures are neither surprising nor distressing. Such failures are not surprising because social phenomena are so complex, and they are not distressing because alternative criteria of evaluation are available for Marx's analysis.

Several other recent commentators on Marx's method have come to similar positions concerning the role of prediction in Marx's system. Roy Bhaskar puts forward much this same objection against the use of predictions in justification in the social sciences. He emphasizes that social systems are causally "open" in the sense that it is not possible to identify a small range of potentially relevant causal conditions for a given set of phenomena. Therefore "social sciences are denied . . . decisive test situations for their theories. This means that the criteria for the rational confirmation and rejection of theories in social science cannot be predictive."[16] He therefore holds that social theories can be tested only through their explanatory power. The shortcoming of this view is the vagueness of the notion of "explanatory power," for on many conceptions of explanation, explanation and prediction are coordinate within a scientific theory. The central point is correct, however: Predictions are substantially less useful as an epistemological standard in social science than they are in typical natural sciences. Similarly Derek Sayer judges that Marx makes "prognoses" rather than predictions. "By a prognosis I mean a hypothesis as to the likely course of future events that, although well-grounded in our analysis of the conditions and mechanisms underpinning present phenomena, cannot be generated out of this analysis by simple deduction. . . . By a prediction, on the other hand, I mean something very much more precise. A prediction is a deduction of what will necessarily follow if (1) certain laws . . . obtain, and (2) requisite antecedent conditions . . . are satisfied."[17] I take Sayer's point to be that Marx cannot make the latter sort of prediction because his system cannot encompass all the relevant laws governing social processes within capitalism. Consequently it is not possible to derive predictions about necessary regularities; at best we can anticipate certain sorts of developments on the basis of abstractive assumptions that may not, however, be satisfied in the future.

Both Bhaskar and Sayer, then, support the basic notion that Marx's system cannot be evaluated through the truth or falsity of its predictive consequences. Their treatments are less than fully satisfactory, however, because they do not provide an alternative account of justification according to which it is possible to use empirical evidence to assess Marx's analysis. And the treatment provided here is intended to provide just such an account.

On the interpretation being offered here, then, testing predictions is only of weak use in empirically evaluating Marx's analysis. But if this is so, how are we to evaluate Marx's construction (including the tendencies just described)? We can do so only on the basis of the factors already discussed: The empirical adequacy of the abstract account Marx gives of the defining features of capitalism and the rigor of his reasoning from these features to their implications for the workings of capitalist society. This suggests an epistemology for social science that turns the H-D model on its head. Rather than evaluating a theory on the basis of its predictions, we evaluate the cogency of the predictions on the basis of (1) the justification available for the initial hypotheses, and (2) the rigor of the argument establishing these tendencies. This model represents a pattern of inference that gives empirical weight to conclusions through fairly long chains of reasoning rather than through direct testing of the conclusions themselves.

Consider an example. Marx argues that the capitalist mode of production tends to economic crises. The argument proceeds by demonstrating institutional sources of economic dislocation and showing that these tend to recur. The sorts of factors Marx identifies include disproportionality between sectors (*Capital II*, pp. 543–44), differences in the rate of turnover in capital goods (*Capital II*, pp. 308–9), fluctuations in the circulation time of commodities (*Capital II*, pp. 356–58), the falling tendency of the rate of profit (*Capital III*, p. 357), and dislocations in the system of credit (*Capital III*, pp. 590–92). But of course other factors may tend to offset crisis, and Marx in fact identifies many of these. So the prediction of a tendency toward economic crisis is only weakly testable by seeing if the capitalist mode of production in fact experiences crises. It is more strongly evaluated by rigorously evaluating the premises and arguments Marx provides. Are the premises of the various arguments empirically supported? And does Marx's reasoning satisfy the appropriate standards of rigor?

These considerations suggest that the justification of Marx's work is simpler than it might seem. To evaluate the analysis, we must ask these questions. Are Marx's premises about the essential features of the system empirically well-founded? Is his reasoning from these sufficiently rigorous? Do we judge that the factors he identifies satisfactorily explain the relevant phenomena? And are there other factors that ought to have been identified along with these? Only secondarily will we examine the long-term consequences of the hypothesis. And this is true because Marx is offering not a general theory but an explanatory analysis of capitalism based on a variety of independent premises. There are various explanations to be evaluated on the evidence and the strength of Marx's arguments; there are basic propositions about capitalist structure to be assessed; but there is no general theory akin to atomic theory that can be evaluated only as a deductive system.

Justification of the Theory of Value

Let us turn finally to Marx's defense of the labor theory of value (LTV). The question we must ask is how the labor theory of value is related to the descriptive and historical evidence just discussed. It will emerge that Marx's arguments in favor of the theory are more conceptual than empirical. This finding supports the conclusion drawn in chapter 3 that the labor theory of value is not an empirical hypothesis.

Marx introduces the theory of value as a necessary consequence of the description of capitalism as a commodity system of production regulated by profit (*Capital I*, pp. 126–31). A commodity system is one that produces goods for exchange, and a commodity system oriented to profits is one in which the exchanges are not conducted for consumption but rather for accumulation of profit. When goods are produced for exchange, however, there develops a socially real distinction between use value and exchange value; the former is the practical and concrete potential the good possesses for the satisfaction of human needs and desires, whereas the latter is the relational property the good has with respect to other commodities – the ratio at which it exchanges with them. Marx argues, therefore, that exchange value must correspond to some socially real property, and he concludes that there is only one property that can do the job. That social property is the quantity of labor time necessary for the production of the good.

Thus we have an argument from the nature of commodity-producing systems in general to the concept of value as socially necessary labor time. A second strand of the argument runs from the institutional goal of profits to the theory of value (*Capital I*, pp. 258–80). Marx argues that profit is impossible in a closed economic system within which goods exchange at their stable exchange value, on the assumption that each producer purchases materials and assembles the product by himself. The only possible source of a surplus is a market disequilibrium in which one producer sells his commodity above its value. This condition cannot give rise to a social net surplus, however, since the gain of the seller is offset by the loss to the buyer. Simple commodity production, therefore – in which independent producers bring their goods to market – is incompatible with profit. Since capitalism rests on the institutional requirement of profit, it follows that there must be some factor the description of simple commodity production fails to take into account: There must be some factor that creates value and therefore allows for a social surplus even in perfect market equilibrium. The labor theory of value fulfills this requirement perfectly; it allows for a single modification of the description of simple commodity production that makes the pursuit of profit possible. That is, if we assume that value is socially necessary labor time, the value of a commodity is a function of the amount of labor time that goes into its production. Therefore one factor *creates* value, namely, human labor. So if we change the description of simple commodity production to include labor power as a commodity (and therefore drop the assumption of independent producers confronting each other in the market), there is a commodity whose use creates more value than its purchase requires. The productive use of labor power by participants in the process of production allow them to achieve a net surplus between the value of the product and the value of the commodities (labor power included) purchased in the course of its production. This is the second strand of argument in favor of the labor theory of value: It permits the resolution of what is otherwise an antinomy in the description of capitalism.

Both of these arguments have a somewhat philosophical air. Marx describes them as though they *entail* the labor theory of value. This they certainly do not do, any more than any body of empirical evidence entails a particular system of description. The labor theory of value is one possible system we can use to describe and unify the empirical regularities and apparent contradictions included in the empirical account of capitalism, but certainly not the only one. When we recall the position taken toward the LTV in chapter 3, however, this philosophical flavor is more understandable. If the LTV is an analytic device rather than a substantive theory, we would not expect to find substantial empirical arguments available to establish its correctness. On this treatment, it is unnecessary to offer any empirical justification for the LTV—any more than it is necessary to justify the calculus empirically before we use it in computing planetary motions. The "justificatory" arguments offered here function well to establish the legitimacy of the instrument of analysis, and that is all the LTV requires.

Allen Wood takes a similar view of these arguments for the labor theory of value: "Marx's 'proof' of the law of value is not taken seriously as such by its author. I think it is best regarded as an expository device. . . . In an important letter to Ludwig Kugelmann, Marx emphasizes that his argument is not a 'proof' of the law of value, and that this law stands in no need of such a proof."[18]

Nonjustificatory Uses of Empirical Evidence

So far in this chapter we have discussed empirical data only in the context of justification. A good deal of descriptive and historical matter in *Capital*, however, does not serve justificatory purposes. The clearest examples of nonjustificatory uses of empirical matter are found in Marx's discussion of the conditions of labor within contemporary capitalism ("The Working Day") and his discussion of industrialization ("The Division of Labour and Manufacture" and "Machinery and Large-Scale Industry").

Marx's treatment in "The Working Day" (*Capital I*, pp. 340–416) is distinguished from other parts of *Capital* by its wealth of empirical detail. Here Marx considers the condition of labor and the length of the workday in British capitalism. He provides extensive evidence for the powerful tendency to extend the workday. This support comes from two sources: the factory inspectors' reports concerning those industries regulated by parliamentary restrictions on the conditions of labor, and public health reports, Children's Employment Commission reports, newspaper accounts, and contemporary descriptions of unregulated industries. This body of evidence supports Marx's chief analytic point in this chapter: that capital does in fact seek to maximize the period of productive labor it receives through the wage contract by extending and intensifying the labor it receives.

Having made this basic theoretical point, Marx could have brought the chapter to a close. However, he does not; he goes on to document the condition of labor in nineteenth-century British society and to analyze the history of the struggle between capital and labor over the state regulation of the length and conditions of the workday. This discussion includes the development of child labor and its eventual regulation; the institution of the Poor Laws and their relation to the capitalist system of production; and the eventual establishment of legal limits on the length and conditions of the workday.

The empirical detail Marx provides in this chapter bears a complex epistemological relation to the abstract analysis of capitalism that is the main work of *Capital*. It goes far beyond what is necessary to support Marx's account. However, this is not Marx's sole aim. Beyond justification, Marx also has a descriptive, almost sociophenomenological aim as well: He intends to provide a detailed portrayal of conditions of life within capitalism that supplements the abstract analysis developed elsewhere in *Capital* and that bears out his condemnation of capitalism as a system of life. A great deal of the material in "The Working Day" functions as an interpretive description of the labor process within capitalism; it is a description that is the more meaningful for resting on an account of the underlying dynamic of the capitalist mode of production and its powerful need to appropriate ever-greater quantities of surplus value. Having made this abstract point elsewhere, Marx turns to a concrete description of the conditions of labor to punctuate and make firm the point in this chapter. Thus the chapter serves both to justify the abstractanalysis of capitalism and to use the analysis as a vantage point from which to organize the available empirical detail.

When we turn our attention to Marx's treatment of industrialization (*Capital I*, chapters 14 and 15), we find that there too his use of empirical detail has two intertwined functions. First, it provides empirical support for Marx's analysis of the capitalist mode of production, and second, it describes certain features of capitalism in substantial detail. Once again Marx is trying both to provide the historical evidence that supports his abstract hypothesis and to document his conclusions with available historical and empirical evidence. The latter practice serves moral and political ends that go beyond Marx's scientific purposes: It documents the humanly destructive features of capitalism. Thus this portion of *Capital* is both justificatory and sociophenomenological because it attempts to provide a concrete historical analysis of capitalism as a mode of production.

In short, much of Marx's use of historical and empirical analysis is both justificatory and documentary. Evidence drawn from the history of capitalism supports the relevant portions of Marx's treatment of the capitalist mode of production insofar as capitalism in fact developed along much the lines that Marx's account predicts. Second, however, much of this data serves as documentation of the reality of the capitalist mode of production in a way that graphically represents Marx's judgment of the destructive aspects of capitalism.

Empiricism and Marx's Method

In chapter 4 it was argued that Marx's method of inquiry is empirical rather than philosophical or dialectical. There we considered a number of textual reasons for judging that Marx pursues a method grounded in detailed empirical and historical research. His critique of speculative philosophy, his statement of the method of historical materialism, and his actual practice in *Capital* all justify the conclusion that he regards empirical investigation as the primary source of knowledge about social phenomena. That conclusion has been further supported in this chapter through an extensive consideration of the variety and forms of empirical evidence Marx adduces in favor of his system; we have found that his methods of inquiry and justification are thoroughly grounded in the requirement that knowledge be supported by relevant empirical evidence. For these reasons I conclude that Marx's method is empirical: He is fundamentally committed to the centrality of empirical research in the construction of substantive knowledge about society. This position finds a good deal of support among recent commentators on Marx's method.

Richard Hudelson formulates the issue of Marx's empiricism rather clearly. He points out that traditional empiricism involved commitment to several now-questionable doctrines about the origins of ideas, as well as to the notion that all substantive beliefs require empirical justification. He then presents a loosened sense of empiricism that does not depend on the theory of ideas contained in the empiricism of Locke and Hume: "What remains [of empiricism] is a commitment to a weakened verificationist theory of meaning understood as the claim that empirical laws must be testable and identification of the methodology of the empirical sciences as the sole knowledge producing method."[19] Hudelson argues that "Marx shares with contemporary empiricists the view that this scientific method is the only avenue to knowledge."[20] His argument is based in large part on the sorts of considerations advanced in chapter 4 concerning Marx's critique of philosophical method.[21] Ernest Mandel also defends the notion that Marx's method is an empirical one in *Late Capitalism*, though on different grounds; he points out Marx's stated commitment throughout his economic writings to the fullest consideration of the empirical evidence. "There is no doubt that Marx considered that the empirical appropriation of the material should precede the analytic process of cognition, just as practical empirical verification should provisionally conclude it."[22] Finally, in his extensive survey of Marx's method, Derek Sayer comes to a similar conclusion. He emphasizes that Marx "remained convinced of the necessity for eschewing speculative construction and starting from premises which . . . 'can be verified in a purely empirical way'."[23]

7

Falsifiability and Idealism

We will conclude our treatment of the empirical standing of Marx's theory of capitalism by considering two critics concerned with this issue. Karl Popper and E. P. Thompson have both leveled serious charges against Marx on grounds relating to justification. Popper argues that Marx's theory is unfalsifiable and therefore not genuinely scientific, and Thompson maintains that Marx tends toward idealism through his excessive confidence in theory. In spite of the ideological distance between these authors, their criticisms have a good deal in common. Both suggest that it is not possible to acquire theoretical knowledge about the course of history or about the development of social processes. And they argue that Marx is insufficiently attentive to empirical and historical evidence. However, I will contend that both have an oversimple conception of empirical reasoning: Popper, because the falsifiability criterion makes scientific research virtually impossible, and Thompson, because he is skeptical about the possibility of theoretical explanation in social science and too confident in the fixity of evidence.

Falsifiability and Adhocness

Karl Popper argues that Marxist theory is an endlessly flexible instrument that can be brought to "account" for every imaginable social state of affairs. Since Marx is prepared to modify his theory in the face of falsifying empirical evidence, the theory is irrefutable and therefore unscientific.[1]

Popper's argument depends on his falsifiability criterion for distinguishing between science and nonscience.

> A theory which is not refutable by any conceivable event is non-scientific. Irrefutability is not a virtue of a theory (as people often think) but a vice.[2]

This criterion is then used to fault Marx's scientific practice (in particular, his readiness to appeal to "countervailing tendencies"):

> The Marxist theory of history, in spite of the serious efforts of some of its founders and followers, ultimately adopted this soothsaying [conventionalist] practice. In some of its earlier formulations . . . their predictions were testable, and in fact falsified. Yet instead of accepting the refutations the followers of Marx re-interpreted both the theory and the evidence in order to make them agree. In this way they rescued the theory from refutation; but they did so at the price of adopting a device which made it irrefutable. They thus gave a "conventionalist twist" to the theory; and by this strategem they destroyed its much advertised claim to scientific status.[3]

In terms of a simple example, Popper's argument may be formulated as follows: Marx's theory of surplus value is initially an empirically significant hypothesis. One of its consequences is that rates of profit are unequal in different industries (because of unequal capital-labor ratios). This is an empirical consequence known to be false; therefore the theory is false. If Marx is unwilling to accept this refutation, preferring instead to refer to other factors that interfere with the operation of the mechanisms specified by the theory, he can do so, but at a cost: he deprives the theory of its empirical content and its scientific standing. Given that Marx routinely suggests modifications of his theory in such circumstances, Popper concludes, his theory of capitalism is unscientific and devoid of empirical content.

Preliminary Issues Concerning Falsification

Before addressing the substance of Popper's argument, it must be noted that many of his "falsifying circumstances" miss the mark because they fall outside the scope of Marx's scientific analysis altogether. Throughout this work I have held that it is necessary to distinguish sharply between Marx's scientific treatment of the capitalist mode of production and other views he offers within his theory of historical materialism. Chapter 1 showed that the scope of the former is limited to an analysis of the logic of the economic structure of capitalism and that it is not committed to broader theses concerning capitalist society at large (e.g., the theories of politics or ideology associated with historical materialism). Marx certainly has strong opinions about the future development of capitalism, and he has a good deal to say about ideology and politics. But none of these areas is included in the core of his science; his *science* is strictly limited to a structural and dynamic analysis of capitalism. Once we draw this line of demarcation, however, we find that most of the facts Popper believes falsify Marx's analysis are actually irrelevant.

Let us consider one example more fully. Popper claims that the occurrence of revolution in Russia and its nonoccurrence in Great Britain falsify Marx's system.[4] Do these considerations count as potential falsifications of Marx's theory? My view is that they do not. First, nothing in *Capital* commits Marx to any position whatsoever on the issue of the possibility of revolution in Russia. The nonindustrial, semifeudal Russia of the nineteenth-century was a radically different social formation from England and fell outside the domain of Marx's scientific investigation altogether. The occurrence of revolution there is irrelevant to the truth of his theory in *Capital*. Marx expressly adopts this position in his correspondence with the editorial board of a Russian periodical. There he is concerned to discourage the application of the argument advanced in *Capital* to the circumstances found in late nineteenth-century Russia: "The chapter on primitive accumulation does not claim to do more than trace the path by which, in Western Europe, the capitalist economic system emerged from the womb of the feudal economic system. . . . [My critic] insists on transforming my historical sketch of the genesis of capitalism in Western Europe into an historico-philosophic theory of the general path of development prescribed by fate to all nations, whatever the historical circumstances in which they find themselves. . . . By studying each of these forms of evolution separately and then comparing them one can easily find the clue to this phenomenon, but one will never arrive there by using as one's master key a general historico-philosophical theory, the supreme virtue of which consists in being supra-historical" (*Correspondence*, pp. 291–93).

Is *Capital* at least committed to the occurrence of socialist revolution in industrialized Europe? It is not. Marx certainly believes that socialist revolution is inevitable in industrialized Europe and that *Capital* supports that belief. Nonetheless this conviction is only a very distant and conditional implication of his theory of capitalism. The more immediate consequences of his theory concern the developmental tendencies of capitalism: changes in property relations and in the techniques of production, concentration and centralization of industrial production, the creation of an industrial reserve army, the recurrence of economic crises, and so on. These are Marx's real "theorems." On the basis of these conclusions Marx offers an argument late in *Capital* to the effect that these tendencies spell the eventual ruin of the capitalist mode of production. That argument, however, cannot be considered a part of the scientific system of *Capital*. It is Marx's somewhat speculative projection of the substantive consequences of his theoretical system rather than a consequence of that system itself. Therefore the failure of revolution in Western Europe undermines only chapter 32 of *Capital*, "The Historical Tendency of Capitalist Accumulation," not the argument of the work as a whole.[5]

Thus this example fails to falsify Marx's system because it describes matters beyond the scope of Marx's scientific analysis. Moreover it is fairly typical of Popper's chief "falsifications" of Marx's system. Consequently his charge against

Marx is weak on its own terms: It fails to offer circumstances that are related to Marx's scientific work in such a way that they could count as falsifiers.

This conclusion does not resolve the issue, however, since we can easily repair Popper's case. In the previous chapter we found that Marx's system *is* committed to some long-range predictions: the falling rate of profit, the intensification of crisis, the "immiseration of labor," and the creation of an industrial reserve army. Marx's response to anomalies of these sorts is to appeal to competing factors that interfere with the basic theoretical mechanisms. Thus Popper's objection can be reformulated in terms of these examples, since none has been unequivocally borne out without need of some qualification. Is Marx's introduction of "countervailing tendencies" to account for certain failures of prediction a "conventionalist twist" that makes the analysis unfalsifiable altogether?

Marx's Use of Countervailing Tendencies

Popper's charge of unfalsifiability finds its strongest ground in Marx's use of "countervailing tendencies." (A countervailing tendency is a previously unknown factor that is hypothesized in order to account for discrepancies between theoretical expectations and observed fact.) Consider this statement of Popper's charge:

> Experience shows that Marx's prophecies were false. But experience can always be explained away. And, indeed, Marx himself, and Engels, began with the elaboration of an auxiliary hypothesis designed to explain why the law of increasing misery does not work as they expected it to do. According to this hypothesis, the tendency toward a falling rate of profit . . . is counteracted by the effects of colonial exploitation.[6]

Appeal to countervailing tendencies is a central part of Marx's notion of a "law of tendency," that builds in the possibility of influences that offset the basic law. However, Popper's falsifiability thesis entails that appeal to such tendencies is itself a conventionalist twist that deprives the construction of empirical content. So let us address this issue directly: Is it irrational to appeal to countervailing tendencies to account for discrepancies between a theoretical analysis of capitalism and its observed characteristics?

We may begin with an example. In volume III of *Capital* Marx confronts an important theoretical problem in political economy: how to account for the falling rate of profit in British industries at midnineteenth-century. Marx explains this tendency on the basis of his theory of surplus value. Profits are equal to surplus labor and therefore proportional to the amount of labor employed. As capitalists introduce more productive techniques to improve profitability, they normally cause the capital-labor ratio to rise. The rate of profit, however, is equal to the ratio of the surplus to total capital employed (wages and machinery), and as wages come to be a smaller share of the total capital, the rate of profit tends to decline. Consequently, Marx's theory of surplus value entails the law of the falling rate of profit (*Capital III*, pp. 317-38). Given the details of the theoretical argument, however, one would expect the rate of profit to fall much more rapidly and constantly than it is observed to do. Whereas orthodox political economy is embarrassed by the mere fact of the fall in the rate of profit, Marx is embarrassed by its relatively slow rate of fall.[7]

Marx's response to this anomaly is to maintain that only a "law of tendency" has been discovered, that may be offset by other factors not yet represented; he then tries to discover the factors that might interfere with the mechanisms specified by the theory of surplus value. "If we consider the enormous development in the productive powers of social labour over the last thirty years alone, compared with all earlier periods, . . . then instead of the problem of explaining the fall in the profit rate, we have the opposite problem of explaining why this fall is not greater or faster. Counteracting influences must be at work, checking and cancelling the effect of the general law and giving it simply the character of a tendency, which is why we have described the fall in the general rate of profit as a tendential fall" (*Capital III*, p. 339).

The factors Marx cites as countervailing tendencies are unobjectionable: He observes that employers have a constant incentive to increase the intensity of the labor process, thereby balancing the decline in the rate of profit; similarly capitalists are induced to depress the wage below the value of labor power; third, capitalists can offset the decline by finding foreign markets on which commodities may be sold above their value (and labor employed below its value); and so forth (*Capital III*, pp. 339-48). All these factors have obvious effects on the fundamental mechanisms of the labor theory of value, and they plainly serve to balance the tendency created by the theoretical mechanisms Marx cites. We may readily see, moreover, the social mechanisms by which they emerge: Capitalists, as they are caught in a situation of generally falling profits, will do what they can to minimize their costs and increase productivity. Introducing these factors into the analysis at a higher level, moreover, seems to be unobjectionable and not a reduction of the empirical content of the theory, since the end result is an increase in the theory's realism.

Popper's attack on countervailing tendencies, however, derives from the following consideration: Given that it is always possible to save a theory from false consequences by referring to interfering factors not accounted for in the theory, does the appeal to such factors not reduce the empirical commitment of the theory? And if the scientist is prepared to make some such appeal in every anomalous case, does he or she not relinquish claim to having provided an empirically significant hypothesis? Is it not reasonable in such a case to conclude that the theory is unfalsifiable by stipulation, and therefore devoid of empirical content?

In other words, is Popper not right in describing recourse to countervailing tendencies as the "conventionalist twist," that establishes the truth of the theory by stipulation?

This conclusion would be justified only if it were impossible to impose limits on the appeal to these tendencies—only, that is, if it were impossible to show how to distinguish between ad hoc and progressive modifications of the theory. And it presupposes that the criteria of scientific rationality attach to formal theories rather than to programs of research. In order to evaluate these charges, however, it will be necessary to consider the issues of anomaly and rational theory change in greater detail.

Anomaly and Theory Change

Popper's falsifiability thesis arises in response to the general problem of anomaly in science. Anomalies—facts or discoveries that appear inconsistent with accepted theory—are found everywhere in the history of science, since scientific inquiry is inherently fallible. If a theory implies some sentence S and S is false, it follows that the theory must be false as well. In such a case the scientist is faced with a range of choices. He or she can reject the theory as a whole; reject some portion of the theory in order to avoid the conclusion S; modify the theory to avoid the conclusion S; or introduce some additional assumption to show how the theory is consistent with "not S." A strict falsificationist would presumably require that we disavow the theory, but this response is both insensitive to actual scientific practice and implausible as a principle of methodology.

When faced with anomaly, the scientist must choose whether to abandon the theory altogether or modify it to make it consistent with the contrary observations. If the theory has a wide range of supporting evidence (aside from the contrary experience), there is a powerful incentive in favor of salvaging the theory, that is, of supplementing it with some further principle restricting the application of its laws, or modifying the laws themselves, to reconcile theory with experience. Ideally the scientist ought to proceed by attempting to locate the source of error in the original theory. Theory modification in the face of contrary evidence should result in a more realistic description of the world, either through the correction of false theoretical principles or through the description of further factors at work that were hitherto unrecognized.

It is possible, however, to modify a theory in ways that do not reflect any additional insight into the real nature of the phenomena in question, but are rather merely mechanical modifications of the theory made to bring it into line with the contrary evidence. Such modifications are common in the history of science; Hempel cites the example of phlogiston theorists under attack by Lavoisier. After Lavoisier's discovery that metals weighed *more* after combustion than prior (thereby apparently falsifying phlogiston theory), some proponents of that hypothesis modified their concept of phlogiston by assuming that it possessed negative weight. This alteration reconciled the phlogiston theory with Lavoisier's evidence; nevertheless it seems on fairly intuitive grounds to be an illegitimate modification. It is "introduced *ad hoc*—i.e., for the sole purpose of saving a hypothesis seriously threatened by adverse evidence; it would not be called for by other findings, and, roughly speaking, it leads to no additional test implications."[8]

The problem of avoiding adhocness by devising a set of methodological standards suitable for governing the modification of hypotheses in light of contrary evidence is a substantial one. As Hempel observes, the clearest judgments of adhocness are made with the benefit of hindsight; what may have been a rational modification given current beliefs is with the benefit of later knowledge a transparent case of ad hoc modification. However, we may advance a rough set of guidelines for the introduction of modifications: "Is the hypothesis proposed just for the purpose of saving some current conception against adverse evidence, or does it also account for other phenomena, does it yield further significant test implications?"[9] Does it contribute to a theory that affords simple explanations of a wide range of phenomena? Does it appear to represent an increased knowledge of the real mechanisms that underlie observable phenomena? Does it merely repeat the evidence already available, or is it amenable to independent tests?[10] These considerations fall far short of a definition of adhocness, and recent work in the philosophy of science has substantially extended these ideas by introducing the notion of a research program.[11]

Postpositivist Treatments of Anomaly

Postpositivist philosophy of science has directed much of its efforts to formulating more adequate standards for modifying theory in the light of anomaly. Its chief insights have resulted from a shift of attention from the level of finished theories to the level of the research program, that is, from the formal laws and principles of a theory to the more encompassing set of presuppositions, methodological commitments, and research interests that guide scientists in the conduct of research and theory formation. The central focus of neopositivist theory of science was the scientific *theory*, conceived ideally as a formal system of axioms and deductive consequences. Neopositivists distinguished between the context of discovery and that of justification, and they argued that only the latter fell within the scope of rational control. This meant that only finished theories could be rationally evaluated, whereas the conduct of research was conceived of as an exercise of pure, unregulated imagination.[12] From this judgment followed falsificationism, verificationism, and various forms of confirmation theory.

The "new" philosophy of science focuses on the "context of discovery"—the assumptions and research goals that guide scientists in their research. Philosophers of science in this area reject the idea that the conduct of research is an unstructured, nonrational process, and they have tried to formulate a theory of the rules that distinguish good scientific research from bad. From this starting point, the "research program" becomes the central interest.[13]

What is a research program? It is the framework of assumptions, experimental procedures, explanatory paradigms, and theoretical principles that guide the conduct of research. Lakatos has provided an especially clear formulation of the concept.[14] First, he argues that any research program possesses a "hard core" of theoretical principles that constitute its central insight into its subject matter. This core is taken as fixed; the "negative heuristic" of the program forbids the scientist from interpreting anomalous results as falsifying this core. Instead, the scientist is directed to construct a "protective belt" of auxiliary assumptions intended to secure the correctness of the theoretical principles at the core, and scientific research becomes an effort to modify or replace the assumptions included in the belt so as to make the core consistent with experimental results. Finally, the research program includes what Lakatos calls a "positive heuristic": a set of principles and assumptions that provide guidance in extending and developing the belt. This conception of scientific inquiry could be summarized in the form of a slogan: Defend and extend! Built into the view is a rejection of falsificationism, for, far from seeking to refute the central theoretical principles, the scientist is directed to defend and extend them as forcefully as possible.

With this fundamentally different starting point, the new philosophers of science have posed a different question for themselves. Rather than the positivists' question – What is the criterion of an empirically adequate theory? – they have asked, What are the features that distinguish a rational and progressive program of research from its contrary? The problem of theory adequacy does not disappear, but it becomes a subordinate concern. This broader approach to empirical rationality lays the emphasis on the degree to which the commitments of the research program successfully direct research productively and suggest empirically adequate theories – rather than on the narrower question of the criterion of empirical adequacy of theories. On this view, empirical rationality is a feature of the program of research rather than the finished theory; theories are tools for understanding empirical phenomena created by the scientist within the context of a framework of methodological and substantive assumptions.

From this research-oriented point of view, falsificationism is an unsound principle of theory choice, since it is an extreme principle that requires the rejection of any theory with false consequences. A more conservative strategy is required, one that allows the scientist to preserve the old theory at minimum cost. On this view, it is generally a reasonable methodological principle to try to formulate a hypothesis that would account for the truth of a theory and the falsehood of one of its consequences S – either by supposing S is really true (i.e., experimental error) or by modifying the theory or by positing some unobserved factor that, together with the theory, predicts "not S." (Consider, for example, the anomalies arising in the Newtonian description of the planetary orbits that led to the subsequent positing of Uranus.) It is reasonable, that is, to take as a research strategy the *maxim of least harm*: to try to produce a reconciliation of theory and observation that requires the least change in the theory. And the general success of scientific theory formation guided by this maxim vindicates the strategy.

The problem with the principle of least harm is that it allows us to stave off rejection of the theory indefinitely; it potentially makes the theory irrefutable. Once we widen our vision from theories to research programs, however, we find that the key problem is not how to keep a theory falsifiable but rather how to impose a set of rational constraints on the principle of least harm: how to avoid ad hoc modifications of theory that fail to advance the theory's empirical power and explanatory adequacy.

Lakatos has discussed this question in detail. His account is not altogether adequate, but it gives an indication of the sort of criterion of adequacy that seems to have some promise of success. On his view, the problem is how to define the notion of a "progressive problem shift," that is, a modification of theory in light of conflicting experience that improves the empirical adequacy of the theory. Lakatos gives a twofold criterion of progressiveness. A modification of theory is *theoretically* progressive if the modification has some excess empirical content over its predecessor, and it is *empirically* progressive if some of this content is corroborated. If the change is not progressive in these senses, the research tradition is in a state of degeneration and ought to be replaced.[15]

This conception of a progressive research tradition may be amplified into a more specific criterion of rational adherence to an empirical theory in the face of anomaly. First, the theory in question must have achieved some empirical success. That is, it must produce empirically adequate explanations of phenomena in areas other than those affected by anomaly; otherwise it would be irrational to remain committed to the theory. And second, the modifications of the theory must themselves be, at least potentially, empirically significant. (1) They must give rise to other consequences besides the range of phenomena they were introduced to explain, and (2) they must be amenable to further investigation. If these conditions obtain, and if independent justification is produced for the new factors, both they and the earlier theory are vindicated.

Application of These Results to Marx's Research Program

Marx's research program satisfies these standards of progressiveness and therefore is an empirically defensible effort in social science. First, the program defines a "hard core" of theoretical principles that appear to constitute an important insight into the workings of the capitalist economy, and as well as a fairly precise tool for further investigation. This core includes the theory of the class nature of capitalism, Marx's account of the defining features of the capitalist economic structure, his analysis of crisis, and so forth. Furthermore this core has had some notable empirical successes. So the research program possesses the basic empirical credentials we found necessary.[16]

Second, the program has been fertile in directing research. Within the area of social science loosely inspired by the Marxian theory of capitalism are to be found indisputably fruitful examples of research. Marxist economic theory has included the contributions of Paul Sweezy, Maurice Dobb, Ernest Mandel, and Michio Morishima; Marxist historiography has produced Eugene Genovese, E. P. Thompson, Albert Soboul, and others; the Marxist theory of politics has given rise to Ralph Miliband, Nicos Poulantzas, and Theda Skocpol; and Marxist sociology has inspired Tom Bottomore and Anthony Giddens, to name only two. This variety of productive efforts indicates that the program has provided the sort of heuristic value expected of research programs.

And finally, the modifications of the theory, and the appeals to countervailing tendencies (the development of the protective belt), that subsequent Marxist social science has witnessed seem by and large to have the progressive properties that Lakatos requires. They have enhanced the empirical power and scope of the theory, and they have been amenable (by and large) to further empirical research (contrary to Popper's assertion that they are merely conventionalist strategies). They have led to a more precise formulation of the economic model; a more detailed model of class stratification; a more finely drawn analysis of the influence of the financial system on the dynamic of capitalist accumulation; and a more precise analysis of the influence of economic structure on the system of education.[17] Marx's program thus gives every appearance of having been both theoretically and empirically progressive, in Lakatos's terms, and it seems to have been fruitful in just the way a successful program of research ought to be.

On the preceding interpretation of theory change in science, then, appeal to countervailing tendencies is neither irrational nor uncontrollable; it is rational to try to save the theory, and we can specify relatively clear criteria of success and failure in the attempt. Appeal to such factors can in principle constitute a progressive research strategy and does not prima facie affect the empirical content of the theory. It may well transpire that the theory is not progressing, and if this is the case, it ultimately must be rejected. But merely to look for ways of accommodating recalcitrant data is not by itself irrational. This is all we need to conclude that the appeal to countervailing tendencies does not by itself reduce the empirical content of the theory. Popper's charge that Marxist social theory is devoid of empirical content is therefore unsupported. Marx's analysis may eventually be shown to be false. Its falsity, however, can be shown only by the long-term failure of detailed research launched in its name and the success of research that contradicts it; it cannot be evaluated in advance of sustained social inquiry.

E. P. Thompson's Critique of *Capital*

E. P. Thompson's recent *The Poverty of Theory*[18] is a spirited attack on Althusser.[19] More generally, however, it is a critique of what Thompson regards as a dark strain of classical Marxism—what he calls Marxist obscurantism (*PT*, p. 168), Marxism as a "self-validating . . . conceptual system" (*PT*, p. 167), and Marxism as "theology" (*PT*, p. 189). I will concern myself not with the justice of Thompson's attack on Althusser[20] but rather with his underlying thesis: that Marx's own work in *Capital* plants the seed for the "Marxist idealism" that he identifies in Althusser and that Marx's effort to construct a scientific theory of the economic structure of capitalism is seriously misguided.

> *Capital* is a mountainous inconsistency. As pure Political Economy it may be faulted for introducing external categories; its laws cannot be verified, and its predictions were wrong. As "history" or as "sociology" it is abstracted to a "model," which has heuristic value, but which follows too obsequiously ahistorical economic laws. (*PT*, p. 65)

Many of Thompson's charges are actually criticisms of naturalism and are sound. As we saw in chapters 1 and 5, however, it is possible to construct a nonnaturalistic interpretation of Marx's analysis. This interpretation allows Marx's account to avoid most of Thompson's charges. Therefore Thompson's case against Marx is weak, and his estimate of the scientific value of *Capital* is unreasonably negative.

Thompson's Chief Criticisms

Thompson assails the strand of Marxism that pursues the aim of creating a scientific theory of capitalism. "It is in the very notion of Marxism as 'Science' that we find the authentic trade-mark of obscurantism" (*PT*, p. 168). As Thompson recognizes, that strand includes Marx himself. He therefore rejects the program Marx defines in *Capital*: the construction of a scientific theory that "reveal[s] the economic law of motion of modern society" (*Capital II*, p. 92). Thompson regards Marx's immersion in political economy as a regrettable distraction from his productive work on historical materialism: "For at least twenty years, Marx had turned aside [from historical materialism] to wrestle with his antagonist, Political Economy. . . . Marx was himself, for a time, trapped within the circuits of capital . . . and . . . he only partly sprung that trap in *Capital*" (*PT*, p. 163). Thus Thompson advances a startling revaluation of Marx's work. He maintains that Marx's most productive effort was expended in his construction of the method and practice of historical materialism and that his economic study of capitalism was work wasted in pursuit of a chimera: the hope for a scientific theory of capitalism.[21]

Thompson's case against *Capital* depends on several chief grounds among which Thompson does not clearly distinguish. First, he criticizes Marx's alleged "reductionism"—his supposed effort to reduce the whole of capitalist society to a set of laws of motion derived from an analysis of the economic structure of capitalism. And second, Thompson attacks Marx's "economism"—his effort to examine capitalism solely as an economic structure without attention to its cultural and political features. These shortcomings, Thompson maintains, lead to a crucial failure on Marx's part: Marx sacrifices attention to the actual historical characteristics of capital*ism* through his attempt to construct a theory of the capital*ist* mode of production. And Thompson's prescription is simple: Marxist studies should abandon the search for a scientific account of capitalism and instead engage in historical research into the particulars of existing capitalist societies.

Reductionism. Let us begin with Thompson's charge that *Capital* is a reductionist work in that Marx intends to explain all important social phenomena within capitalist society on the basis of facts about the economic structure of capitalism. Against this project, Thompson holds that "a unitary knowledge of society . . . cannot be won from a 'science' which . . . isolates certain kinds of activity only for study, and provides no categories for others" (*PT*, p. 65). Here Thompson plainly presupposes that Marx intends to provide a "unitary knowledge of society" in *Capital*—a single, unified theory that permits the theoretician to organize and explain the full empirical reality of capitalist society on the basis of economic facts alone. Thompson holds that this goal is unattainable because noneconomic phenomena (e.g., political, ideological, and cultural factors) cannot be reduced to economic categories. Therefore Marx's exclusive analysis of economic phenomena makes it impossible for him to give a full-blooded account of capitalist society.

On Thompson's account, the most an economic analysis of capitalism can provide is an abstract model of the operation of the economic structure of the system, but this model does not permit us to derive consequences for the workings of noneconomic aspects. "The problem . . . is to move from the circuits of capital to capital*ism*; from a highly-conceptualized and abstracted mode of production" (*PT*, p. 163). Capital*ism* is a social whole consisting of irreducibly diverse phenomena, whereas the capitalist mode of production is a homogeneously economic abstract entity. Political economy can at best inform us about the laws of motion of the latter; it cannot provide detailed information about the former. Thus Thompson maintains that Marx's exclusive attention to economic categories guarantees that he will not be able to account for the bulk of capitalist society.

This charge assumes that Marx's aim in *Capital* is to provide a unified analysis of the whole of capitalist society, but this assumption is incorrect. Instead (as we saw in detail in chapter 2), Marx imposes sharp limits on the scope of his account, and he leaves it for further research to establish the explanatory re-

lations that obtain between his economic results and "regional" theories—
theories of politics, ideology, and the like.[22] *Capital* is an elaboration of the economic structure of capitalism (i.e., the social relations of production on which
that order rests) and of the dynamic of development that structure imposes on
the forces of production. Its purpose is to "lay bare the *economic* laws of motion"
of the capitalist mode of production. It is therefore silent about the noneconomic
features of capitalist society. In particular, Marx does not assert that it is possible to deduce the structure of the capitalist state or capitalist ideology from his
economic theory, and he does not describe *Capital* as a full theory of capitalist
society. Rather it is to be the basis for other empirical investigations. (Indeed,
it is difficult to think of a Marxist social scientist who puts forward a reductionist
theory of ideology or culture that would be subject to the criticisms Thompson
formulates in this context; Hindess and Hirst are exceptional in their willingness
to assert that a pure theory of the mode of production can permit the derivation
of superstructural consequences.)[23]

Thus the scope of Marx's research in *Capital* is carefully limited. *Capital*
does not contain developed theories of politics or ideology, and its central arguments do not depend on assumptions about politics or ideology derived from
historical materialism more generally. And this in turn demonstrates that Marx
does not attempt to present a closed theoretical system capable of explaining the
whole of capitalist society, nor does he try to reduce capitalist society to his analysis of the economic structure of capitalism. Instead, he gives us a specialized
account of the logic of that structure and proposes that the results of this research
will have a central explanatory role within theories of the capitalist state, capitalist ideology, and so forth. But these latter theories can be discovered only
through focused and rigorous empirical investigation of the relevant phenomena
of capitalist society.

This point is further emphasized when we recall that Marx allows that different capitalist social formations can have substantially different characteristics.
They can have major structural features in common but differ in important ways
(e.g., rate of economic development, intensity of class struggle, form of politics). Both aspects of this point are expressed quite plainly in the preface to the
first edition of *Capital*. There Marx affirms that distinct capitalist systems will
show the same process of development:

> It is a question of these laws [of capitalist production] themselves, of these
> tendencies winning their way through and working themselves out with
> iron necessity. The country that is more developed industrially only
> shows, to the less developed, the image of its own future. (*Capital II*, p.
> 91)

But in the same passage he states that this process may take different forms de-
pending on (among other things) "the degree of development of the working

class itself" (*Capital II*, p. 92). If Marx believes that the theory of capital is compatible with a variety of state forms and distinct social formations, he demonstrably does *not* believe that the characteristics of the mode of production determine all other aspects of the social formation. Putting the point more plainly, Marx cannot believe that the capitalist mode of production is the cause, and the social formation the simple effect (as Thompson suggests; *PT*, p. 61).

We can accept Thompson's basic point: *If* Marx believed it were possible to provide a single general theory that would account for every aspect of British capitalism (political, religious, economic, ideological, etc.), Thompson would be perfectly right in rejecting this. Such a program would depend on unreasonably deterministic assumptions about social processes and overoptimistic estimates of the degree of precision available to social science. However, these arguments show that Marx does not intend for *Capital* to provide a deductive theory that reproduces the entire rich texture of capitalist society. We have seen that his purposes are quite limited: to provide a theory of the *economic* laws of motion of the capitalist mode of production. It remains for historical materialism more generally to work out the institutional logic these laws impose on capitalist society as a whole. As a result Thompson's charge that *Capital* cannot in principle provide a full theoretical explanation of noneconomic phenomena misses the mark altogether; Marx does not intend to provide such a theory in *Capital*.

Such arguments emphasize the limitations Marx imposed on his economic research. It is plain, however, that he believes that his analysis of the economic structure of capitalism will play a central role in future explanations of superstructural phenomena—politics, ideology, and so forth. This is indeed the central point of the theory of historical materialism. Marx does believe that the economic structure of capitalism has an identifiable dynamic, and it has specific connections with various noneconomic formations. Future research within the theory of historical materialism, therefore, can be expected to show how features of the economic structure influence and constrain features of the superstructure. Thompson appears to reject even this modest goal on the ground that it is not possible to reduce noneconomic phenomena to economic categories. This position reflects an untenable assumption, however: that explanatory primacy entails descriptive adequacy.

> Political Economy, including Marx's "anti" structure, had no terms—had deliberately, and for the purposes of its analytical science, *excluded* the terms—which become, immediately, essential if we are to comprehend societies and histories. . . . It has no terms for other areas of consciousness. (*PT*, p. 164)

The implicit criticism here is that cultural and political phenomena cannot be characterized in terms of the categories of political economy, and from this

Thompson concludes that it is not possible for facts about economic phenomena to *explain* noneconomic phenomena.

Examples from other areas of science refute this assumption, however. That the concepts of a given scientific theory do not suffice as a descriptive framework for a given range of phenomena does not establish that the theory is not *explanatorily* fundamental to those phenomena. The pragmatic standards a framework of description must satisfy are largely independent from those that establish explanatory primacy. For example, the concepts of molecular biology are insufficient as a descriptive system for animal taxonomy, and yet the theory of the biochemistry of the gene is central in the *explanation* of the facts reported by the taxonomist.[24] Likewise the vocabulary of quantum mechanics is silent about solubility, boiling point, hardness, and the rest of the language of descriptive chemistry, and yet quantum mechanics is fundamental to the explanation of chemical phenomena. In each case one level of phenomena is thought to be explained by a set of factors that are more fundamental, without it being possible to *reduce* the one level to the other. In the same way Marx is free to admit that the phenomena of religion or politics require their own descriptive frameworks, but still hold that the theory of the economic structure provides the explanatory basis for understanding these features. The point that the concepts of the theory of the economic structure are inadequate for describing the phenomena of religion is irrelevant to this claim.

Isolation of the "Economic." This response brings us to Thompson's second charge, however, namely, his criticism of "economism." For the foregoing interpretation of Marx's purposes in *Capital* seems to provide strong support for Thompson's accusation that Marx intends to isolate the "economic" from other aspects of capitalist society. Thompson challenges the scientific appropriateness of this effort, however. "Marx*ism* has for decades been suffering from a wasting disease of vulgar economism. . . . Now it has swiftly passed into a last delirium of idealism" (*PT*, p. 168). Against this effort to separate the economic from law, culture, and ideology, and to treat the latter within the "superstructure," Thompson puts forward his own experience as a working social historian: "I found that law did not keep politely to a 'level' but was at every bloody level; it was imbricated within the mode of production and productive relations themselves; . . . and it was simultaneously present in the philosophy of Locke" (*PT*, p. 96).

Thompson's point appears to be that there is no stable and causally closed range of phenomena that can be singled out as economic; social processes are inextricably political, cultural, axiological, ideological—*and* economic. It is therefore impossible to identify the economic as a proper subject for scientific investigation, since it is not generally possible to explain economic phenomena in purely economic terms. It is necessary to introduce cultural, ideological, or political factors into the explanation as well. The challenge scientific Marxism must confront along these lines, therefore, can be put as follows: Is it possible to identify the "economic structure" of capitalism as a system that obeys its own laws and that can be scientifically investigated in (relative) isolation from cultural, political, or ideological factors?

Before answering this charge we need to be more specific about the sense in which Marx isolates the economic. Thompson presupposes that Marx selects the purely economic categories of political economy for logical analysis—value, profit, rent, and wages, to name a few. This interpretation is incorrect, however. As we saw in chapter 1, what Marx isolates is the capitalist *economic structure*. But this is defined as a set of social relations of production, or property relations—not simply as abstract economic categories. And these relations cannot be described solely in terms of the categories of pure political economy; they require information about the relations of power and authority that establish control and enjoyment of the productive forces of capitalism—machines, factories, mines, land, railroads, and so on.[25] This range of information about capitalist society is selective because it excludes information about culture, politics, law, and the like. But it is not narrowly economic; it depends on detailed information about the social relations that define the capitalist economic structure.

This sense of the economic is substantially broader than that Thompson attributes to Marx in that it relies on an analysis of certain fundamental features of the social organization of capitalism and not merely on ahistorical economic categories. The economy Marx isolates in *Capital* is historically and empirically specific, and his account of this economy reflects his view of the specific social relations that distinguish capitalism from other economies (classical slavery, feudalism, etc.).

It was reasonable for Marx to attempt to set apart the economic in this less restrictive sense for at least two reasons. First, it is a commonplace that scientific research must be focused through a fairly strict definition of the scope of investigation. It is only by defining a manageably narrow range of phenomena for study that the sort of rigor and detail needed in empirical science is possible. And to identify an area for detailed empirical research is not to commit oneself to the proposition that the resulting theory will be fully general for other domains, or that phenomena in other domains may be reduced to the laws governing the area under current investigation. Thus Marx's effort to single out the economic structure of capitalism for specialized attention does *not* entail his believing that an economic theory suffices to replace all other investigations. Instead it reflects the simple fact about empirical research that the scope of scientific investigation always must be limited. Thus it is open to Marx to hold that capitalist society depends on irreducible elements of noneconomic structure (e.g., political institutions) and even that the economic structure itself requires these institutions (e.g., in the form of legal guarantees of property rights), but still to confine his attention to the strictly economic process of the system. Marx properly chooses to leave noneconomic phenomena to other specialists working within the general framework of historical materialism.

To be sure, Marx concentrates his attention on the economic structure of capitalism because he believes it to be explanatorily fundamental to capitalist society as a whole; historical materialism maintains that the economic structure imposes a logic on noneconomic institutions through the limits and conditions it creates. For example, the capitalist economic structure creates a privileged class (the bourgeoisie) that is empowered to defend its interests politically; consequently the legislative process within a capitalist democracy engenders a system of law that is in substantial harmony with the interests of the capitalist class.[26] Likewise, legislative efforts that are significantly contrary to the needs of the capitalist economic structure run substantial risk of inducing economic crisis, thereby undermining their own stability.[27] This conditioning of the noneconomic by the economic leads to the result that any explanation of the noneconomic institution must draw on the theory of the economic structure. This premise defines the research program of historical materialism as presented in *The German Ideology* and elsewhere, and *Capital* falls within that broad program of research. But though this assumption explains why Marx chooses to expend so much effort in constructing an economic theory, it does not lead him to the unjustifiable conclusion that this economic theory is sufficient all by itself to account for the whole of capitalist society. It remains for historians to utilize the economic theory as well as the other tools of their craft to devise explanations of particular features of capitalist society.

Thus Marx's view of the relation between the economic structure and noneconomic institutions is quite consistent with the process of historical explanation Thompson endorses: the historical process "as the exerting of pressures, as a logic of process within a larger (and sometimes countervailing) process" (*PT*, p. 163). In concentrating his attention on the categories of political economy, Marx presupposes that the needs of the economic structure are central among the "pressures" that shape capitalist society. He is at liberty, however, to acknowledge the autonomous importance of non-economic factors as well. In this connection Thompson has assigned responsibility to the wrong culprit: It is not *Capital* but Marx's writings on historical materialism more generally that tend to assert an exclusive primacy for the economic structure. Consider, for example, the strong version of the thesis of historical materialism in the preface to *A Contribution to the Critique of Political Economy*: "The totality of these relations of production constitutes the economic structure of society, the real foundation, on which arises a legal and political superstructure and to which correspond definite forms of social consciousness" (*EW*, p. 425). By contrast, *Capital* is largely silent on the question of the relations between economic structure and superstructure.

Once again, my assessment of Thompson's criticisms is that they overstate

Marx's theoretical ambitions. It is not Marx's intention to derive the pattern of development of capitalist society as a whole from a set of economic premises. Rather he seeks to isolate the defining features of the capitalist economic structure—the property relations and incentives that constitute the capitalist economy—and to derive from the resulting system a set of laws of development of the capitalist economy. But—on my view of Marx's purposes—it is apparent (1) that these laws may be modified or offset by other noneconomic tendencies within capitalist society (e.g., the workings of religion or politics), and (2) that non-economic tendencies cannot be reduced to economic categories.

Assessment

This account shows that Thompson's criticisms of *Capital* are largely unfounded. Marx does not believe that his theory of the economic structure of capitalism is a full theory of capitalism, and he does not claim that the theory permits a deduction of the full empirical detail of capitalism. Nor does he claim that every aspect of politics, religion, ideology, and so forth, may be deduced from the theory. He *does* claim that an account of the economic structure of capitalism is fundamental to any explanation of the workings of noneconomic institutions and that such institutions commonly may be understood as satisfying the functional requirements of the economic structure. Moreover, we have seen that it is possible to provide a nonreductionist account of explanation that shows this aim to be reasonable.

We also have seen that the scientific content of *Capital* may be identified fairly readily. Marx provides an abstract description of important features of the economic structure of capitalism: the property relations that define the use and control of the productive forces and the institutionally defined goals that establish the purpose of capitalist production. And throughout three volumes of *Capital* Marx attempts to work out various particular institutional consequences of these social relations. His arguments typically take the form of "institutional-logic" explanations in which he tries to determine the collective consequences of the economic activity of individuals within the context of the incentives, prohibitions, and opportunities defined by these social relations for representative individuals within the economic structure. Finally, the fruits of this form of analysis are to be found in the "laws of motion" of the capitalist economy at which Marx arrives: the laws of accumulation of capital (the creation of an industrial reserve army, the tendencies toward concentration and centralization, the tendency toward technical innovation); Marx's theory of economic crises; and his derivation of the law of the falling tendency in the rate of profit.

These points show that Thompson's case for the sterility of Marx's economic analysis is a weak one, and his criticisms of *Capital* are largely unfounded. Marx is not guilty of the latent idealism Thompson attributes to him. Instead, his economics represents a focused, rigorous study of the capitalist economy; the work has borne significant fruit; and it is appropriately supported by empirical data. Since skepticism on these points was Thompson's chief reason for denigrating Marx's economic program, we may reject as well his judgment that Marx's greatest contribution lies in his formulation of the theory of historical materialism. On the contrary, Marx's great achievement is his analysis of the capitalist economic structure. Properly understood, that account provides a basis for the explanation of other aspects of capitalist society. At the same time Marx's position leaves full scope for the sort of historical investigation of the particular features of noneconomic institutions Thompson favors.

Conclusion

Most broadly, I have attempted two related projects in this book. First, I have sought to rediscover Marx as a social scientist – not a social critic, not a revolutionary, but an empirical social scientist. (This is not to deny, of course, that he was also a social critic or a revolutionary, but only to insist that his scientific achievements are noteworthy quite independently from his other contributions.) I have undertaken to explore the most important characteristics of Marx's scientific work – the implicit theory of science defined by his empirical and theoretical work. And I have argued for an interpretation of *Capital* that represents Marx's system as a rigorous and empirically controlled investigation of the capitalist economy.

Second, I hope to have shown that Marx's contributions to social science are not primarily methodological or philosophical, but are rather empirical and explanatory. In disagreement with many orthodox Marxists (e.g., Lukács), there is no uniquely "Marxist" method of social science; rather, Marx's methods of inquiry, reasoning, and justification are in substantial continuity with the larger traditions of empirical scientific investigation. In particular, Marx's theory of science does not depend upon esoteric philosophical systems – whether Hegel's dialectics or Engels's "dialectical materialism" – and it does not represent a decisive break with "bourgeois" social science. On the contrary, Marx's scientific practice stands in strong continuity with orthodox social science. What distinguishes his science from that of Pareto, Mill, or Weber is a set of substantive theoretical disagreements about the fundamental causes of various sorts of social phenomena. Marx's basic insight is that the social relations of production – the property relations – have profound implications for the organization and development of the economy of a society, as well as for higher-level social institutions and relations. It is at the level of Marx's specific analytic insights into the workings of capitalism that his real significance as a social scientist must be located.

Much of the discussion throughout this book suggests that contemporary social science has much to learn from Marx's analysis of capitalism, but these are
lessons to be learned primarily from his practice, not from his pronouncements
as a methodologist. Central among these are his emphasis on explanation and
his rejection of simple inductivism. His commitment to the primacy of empirical
evidence in the assessment of scientific belief is plain, but equally plain is his
recognition that insight into complex social phenomena cannot be gained
through mechanical observation or inductivist generalization. Rather, explanation requires discovering the underlying causes of social phenomena. In this respect too Marx's scientific treatment of capitalism stands as a strong paradigm
of one mode of social knowledge.

Let us briefly review the main line of argument in the preceding chapters.
In chapter 1 we found that Marx's scientific study of capitalism does not take
the form of a comprehensive, unified theory of capitalism as a strictly law-
governed system. Instead Marx's analysis can be represented as a family of
related efforts. First, Marx offers abstractive descriptions of some of the social
relations and institutions he held to define the basic dynamic of capitalism. Central among these premises are his account of the social relations of production
of the capitalist system, his description of the process of class formation, and
his account of the structural characteristics of a system of commodity production
for profit. Following Marx's practice, we can refer to these features as
representing the "inner physiology" of the capitalist system; they define the economic structure of capitalism.

Second, Marx offers a large number of analytic arguments throughout *Capital* in which he attempted to work out the consequences of specific features of
the inner physiology for the organization and development of the capitalist economy as a whole. We described these arguments as efforts to discover the institutional logic created by various parts of the inner physiology of the system. Chief
among Marx's results using this type of reasoning are the tendencies toward
rapid technological innovation, concentration and centralization of productive
wealth, a fall in the rate of profit, economic crisis, and the creation of an industrial reserve army. These include both technical characteristics of development
of the capitalist economy (e.g., the falling rate of profit or the formation of
larger units of production) and its development as a class system (e.g., the
solidification of both major classes through centralization of wealth, on the one
hand, and the formation of an educated, organized working class, on the other).
Marx believes that such tendencies represent the main features of development
of capitalism, and that they follow in a comprehensible fashion from his account
of the character of the inner physiology of the system. A conclusion to be drawn
from this treatment of Marx's reasoning is that the "laws of motion" he discovers
are not akin to laws of nature. They are rather "laws of tendency" whose neces-

sity derives only from the collective consequences of large numbers of individuals acting in more or less rational ways in pursuit of their interests.

In chapter 2 we saw that Marx's study of capitalism proceeds within the context of a larger program of research: his theory of historical materialism. In particular, the fundamental insight of historical materialism—the analysis of a society as a system of forces and relations of production—constitutes the starting point of Marx's account of capitalism as well. In *Capital* he examines the character of capitalist property relations in detail and attempts to work out their long-term institutional consequences for the capitalist economy. For this reason it is important to understand the chief concepts and explanatory models of historical materialism if we are to understand the purpose of Marx's research in *Capital*. Chapter 2 described the main ideas of historical materialism and considered several outstanding controversies in current discussions of historical materialism: the degree to which Marx is a technological determinist, the character of "determination" he means to assert between economic and noneconomic institutions, and the role of class in explanations within historical materialism. It was also argued in this chapter that Marx's economic research is logically independent of historical materialism: *Capital* does not depend on the truth of the theories of ideology, revolution, politics, and the like, that lie at the heart of historical materialism. Thus *Capital* is inspired by historical materialism, but its truth or falsity depends on more specialized evidence drawn from the phenomena of capitalism.

Having identified "institutional-logic" explanations as the central explanatory tool in Marx's system, it is necessary to have a clearer conception of their character. This was the task of chapter 5. There we discovered that Marx's chief explanations take the form of reasoning from the hypothetical situation of a representative class actor (usually the capitalist) to a conclusion concerning the most rational strategy in those circumstances; he then proceeds from the individual's best strategy to the collective consequences for the capitalist system as a whole of the generalization of that strategy. A paradigm example of this is found in Marx's analysis of the impulse toward technological innovation within capitalism. Each capitalist wants to increase the rate of profit of his own firm; each therefore seeks out cost-cutting innovations; and the net result is that—industry by industry, firm by firm—the capitalist economy is driven toward ever-higher levels of technological sophistication and labor productivity. In this chapter we found, moreover, that this sort of account pays special attention to the need for "microfoundations" for Marxian explanations. For Marx recognizes that it is not enough to show that a given strategy would serve the interests of a class as a whole; it is also necessary to show that this strategy serves specific individual interests sufficiently to induce individuals to pursue the strategy. Thus on this account Marx's own explanatory practices are logically similar to work now being done on the microfoundations of Marxism by Roemer, Brenner, Elster, and others.

Central to many of Marx's institutional-logic explanations is his use of the tools of classical political economy. It is this feature that identifies Marx's treatment as an economic analysis and not simply a sociological description of capitalist institutions. Accordingly, in chapter 3 we considered the character of Marx's economic reasoning. There we found that the core of his economics concerns the process of surplus creation and surplus extraction. The labor theory of value is Marx's analytic device for representing the capitalist economy as a flow of value and surplus value. But it was shown in several important cases that it is possible to reproduce Marx's economic explanations without making essential use of the labor theory of value. Thus it was argued that the labor theory of value may be reduced in importance in Marx's system without fundamentally overturning the essential point: that capitalism is an exploitative system in that the social surplus is expropriated by the capitalist class from the class of immediate producers.

Critical to assessing the claim of scientific standing for Marx's study of capitalism must be the empirical status of his account. Consequently in the closing two chapters we turned to Marx's use of empirical data. In chapter 6 we found that he makes substantial and rigorous use of empirical data, both in his method of inquiry and his method of justification. Indeed, this is one of the features that distinguishes *Capital* from Marx's writings in the theory of historical materialism. We also found, however, that his use of evidence does not conform to the deductivist model of confirmation. In particular, Marx's system is not — and should not be — tested on the basis of the success or failure of its predictions. This is because his predictions are soft ones based on laws of tendency. Instead the justificatory force of Marx's construction depends upon the empirical accuracy of his description of the basic institutions of capitalism (the inner physiology) and the rigor of his reasoning from this account to specific predictive consequences. Justification thus flows from the basic premises to the predictive outcomes, rather than the reverse. In chapter 7 we considered charges by Karl Popper and E. P. Thompson that Marx's empirical practices were flawed — that his system is unfalsifiable and implicitly idealist. There it was found that both Popper and Thompson misconstrue the nature of empirical justification in social science and that Marx's account is innocent of the charges both authors bring against it.

These arguments all depend chiefly on Marx's actual explanatory and empirical practices. In chapter 4 we turned to a number of issues arising from Marx's explicit methodological pronouncements rather than from his practice as a scientist. We considered his essentialist theory of science, his abstractive method of inquiry, and his purportedly "dialectical" method of theory formation. It was argued that the first two views represent innocuous theories of scientific explanation and investigation that do not, however, provide a great deal of guidance in the actual conduct of scientific research. The last issue—the status of dialectics in Marx's system—was judged to be more serious: If Marx's account is fundamentally dialectical, it is unlikely to be empirical. However, it was argued that Marx's critique of speculative philosophy prevents him from using a dialectical method, and further it was shown that his infrequent use of dialectical language in *Capital* can be explained in a way that leaves his method fundamentally empirical. Thus Marx's science does not make use of a dialectical method (although it does rely on some substantive assumptions about social change that can be described as "dialectical"). Chapter 4 concluded that Marx's explicit theory of science is largely consistent with current ideas in the philosophy of science concerning the nature of theory formation, the use of empirical evidence, and the nature of scientific inquiry. We described this family of views as a form of "Galilean empiricism."

As this brief survey indicates, much of Marx's science is rather dissimilar from the constructions of natural science: Marx's account is not a comprehensive, unified theory, it is not confirmed through test of its predictions, its explanations do not rely on laws of nature, and its hypotheses are generally couched in descriptive or observational terms. Do these dissimilarities from the natural sciences suggest that Marx's construction is less worthy of our esteem as a work of science? This judgment would be warranted only if we regarded naturalistic theories as the ideal for all scientific knowledge. But the constructions of social science generally do not conform to that standard, and this failure casts doubt on the standard itself rather than on the credibility of social science. Philosophers of social science must recognize the diversity that exists among scientific forms of scientific knowledge. Thus to conclude that *Capital* is not a naturalistic theory is not to conclude that it is unscientific. Indeed, throughout this book we have encountered evidence to the contrary—that Marx's work in *Capital* represents a substantial scientific achievement, a rigorous scientific analysis distinct both from theories in natural science and from the research hypotheses of historical materialism.

This conclusion rests on a varied set of grounds. First, Marx's explanatory goals are fairly precisely defined in *Capital*, and it is reasonable to suppose that scientific investigation of this region will bear fruit. It is reasonable to expect that there is a logic imposed on social life by the economic institutions of a capitalist economy and that this logic may be scientifically investigated. Second, the interpretation given here of Marx's analysis of capitalism has emphasized the limits Marx imposes on its scope. Marx's account of capitalism is limited to discovering the logic of the capitalist economy (although historical materialism gives him reason to suppose that this logic is highly relevant to noneconomic institutions as well). Again, this precision is welcome from a scientific point of view. Finally, Marx's examination of capitalism consists of a set of hypotheses that are detailed enough to permit scientific application. In sharp contrast to his theory of ideology, Marx provides a thorough analysis of the defining features of capitalism that promises to have predictive and explanatory power.

Thus the account of the capitalist mode of production represented here is a serious scientific analysis in this sense: It embodies a set of hypotheses about the defining features of the capitalist mode of production, in contrast to feudalism, slavery, and other systems; these hypotheses are sufficiently precise to permit Marx to draw inferences about the capitalist mode of production (for example, that productivity will tend to rise dramatically within capitalism); and they are intended to be *explanatory* – they are asserted to be the features of the mode of production that give rise to its basic characteristics of organization and development.

Furthermore, arguments offered in the final two chapters uphold the conclusion that *Capital* is an empirically grounded work of science. It provides an appropriate range of evidence – historical, sociological, and economic – in support of the main lines of Marx's analysis. Moreover, Marx's work has been a fertile guide to further research; his model of the capitalist economy has inspired important work in historical research, sociological analysis, and economic theory. *Capital* therefore represents a strong example of a work of social science, both in terms of its theoretical clarity and empirical justification, and its status as a guide for further research.

Throughout this work I have examined the logical features of Marx's social science, not its correctness as an analysis of capitalism. In discussing Marx's use of empirical evidence, for example, I have not been concerned to discover whether the available evidence confirms or falsifies Marx's account, but rather the logical question, namely, whether Marx uses evidence in such a way as to permit him to empirically evaluate his account. Thus my primary endeavor has been to examine Marx's practice as a scientist and to determine whether his efforts at explanation, inquiry, and justification are reasonable ones within social science. It may be appropriate in closing, however, to offer a view of the status of *Capital* as a body of theory about a social and economic system that continues to dominate our lives in the West. Is *Capital* still capable of offering scientific insights into the nature of twentieth-century capitalism?

There is a sense in which Marx's own views would make him suspicious of the claim that an investigation of the social relations of production of nineteenth-century capitalism should remain valid for the social system that emerges from that mode of production over a century later. For Marx is insistent on the historical specificity of the relations that define any mode of production. He raises this point in connection with cross-modal judgments of timelessness (for example, the idea that precapitalist modes of production must "really" have been based on bourgeois exchange relations). But the point is equally valid in application to the development of a single mode of production over time. To the extent that the social relations of production that define twentieth-century capitalism are significantly different from those that defined nineteenth-century capitalism, Marx's analysis must be modified before it can offer relevant commentary on the present.

There are unmistakable differences between capitalist property relations in 1850 and in the mid-1980s. On the side of capital, at least these changes have occurred: the accelerated separation of ownership and management, the increasing role of finance and credit within capitalist enterprises, the creation of the modern multinational corporation as the basic unit of capital, and the increased involvement of the state in the affairs of capitalist enterprises. Changes have emerged on the side of labor power as well: increasing government regulation of work conditions, the shift from industrial to service employment, the creation of effective units of organized labor in all capitalist countries, the rise of mass-based socialist parties with proletarian support in Western Europe, and the emergence of much more extensive social welfare systems in all capitalist systems. All these factors potentially may influence the dynamics of modern capitalism, and they all were of only minor importance in the economic structure Marx investigated.

At the same time there are substantial continuities between nineteenth-century and twentieth-century capitalism. (It is these continuities that justify our identification of the modern American or Western European economies as "capitalist.") The fundamental requirements of capitalist property still exist, namely, the effective separation between a minority class that owns and controls the vast majority of all productive wealth and a majority class that possesses no productive wealth and is obliged to sell its labor power. Capitalism remains a system of class power and privilege—witness the uninterrupted power and influence of the minority class that owns and directs the productive wealth in each capitalist nation. Capitalism remains a system in which class power and privilege derive from ownership of wealth—witness the sharp inequalities of wealth and income that persist to the present day. Moreover capitalism continues to depend on the accumulation of capital, and it continues to reflect a deep conflict of interest between owners and workers in the productive process. Finally—in Marx's technical meaning of the term—capitalism remains an exploitative system: The social surplus is still expropriated from the class of immediate producers by the class of owner of productive wealth.

Given these important similarities and differences between the property relations of nineteenth- and twentieth-century capitalism, it becomes a problem of continuing research—of the sort Marx provided so extensively in *Capital*—to determine whether the fundamental dynamic of contemporary capitalism should be predicted to resemble that of nineteenth-century capitalism. Only detailed empirical and theoretical analysis will permit us to determine whether the continuities are sufficiently fundamental to offset the alterations introduced by changes in the social relations of property and class. For we have seen that Marx's arguments for the "laws of motion" of capitalism depend essentially on assumptions about the details of capitalist relations of production, and those relations have not remained fixed.

This finding suggests that the application of the findings of *Capital* to contemporary capitalism must be somewhat tentative; it is surely *not* possible to derive particular laws of motion of contemporary capitalism from Marx's analysis alone. Rather, Marxist social scientists and political economists must provide the sort of detailed account of modern property relations and economic institutions that Marx provides for nineteenth-century relations and institutions. This is not to say that Marxist social science must begin de novo. The continuities between modern capitalism and nineteenth-century capitalism are crucial to understanding modern capitalist phenomena, and Marx's analysis of those basic features of capitalism remains profoundly illuminating. But it is necessary to supplement, modify, and extend his account to draw particular conclusions about the course of modern capitalism.

Significantly, contemporary Marxist social science conforms to this view of the relevance of *Capital* today. Thus Marxist political economists have put forward detailed studies of modern capitalist relations of production—e.g. Ernest Mandel's *Late Capitalism* and James O'Connor's *The Fiscal Crisis of the State*. Other Marxist social scientists have offered analyses of particular postcapitalist modes of production—Rudolph Bahro's *The Alternative in Eastern Europe* or Donald Hodges's *The Bureaucratization of Socialism*, to name two. And Marxist political sociologists have refined and extended his treatment of class, property, and politics, for instance, Erik Olin Wright's *Class, Crisis and the State* and Ralph Miliband's *The State in Capitalist Society*. What these works have in common is not pious deference to Marx's texts—in *Capital* or elsewhere. Rather, they are unified in being vigorous attempts, using extensive contemporary data, to offer theoretical accounts of modern social institutions within a framework of analysis that is greatly indebted to Marx's treatment of capitalism. Contemporary Marxist social science is rooted in Marx's insights, but it is not confined to his conclusions or to the particular features he singled out for fine-grained analysis. Thus Marx's *Capital* established a research program for twentieth-century social science, and it is a program that has borne fine fruit indeed.

Notes

Notes

Introduction

1. For an extensive consideration of some of Marx's predictions see Fred Gottheil, *Marx's Economic Predictions* (Evanston, Ill.: Northwestern University Press, 1966). Ernest Mandel offers a thoughtful treatment of Marx's predictions in *Late Capitalism* (London: Verso, 1978), pp. 13–43.

2. Naturally there are significant exceptions to this generalization, e.g., Louis Althusser and Etienne Balibar, *Reading Capital*, trans. Ben Brewster (London: New Left Books, 1970), Derek Sayer, *Marx's Method* (Atlantic Highlands, N.J.: Humanities, 1979), Russell Keat and John Urry, *Social Theory as Science* (London: Routledge & Kegan Paul, 1975), and Laird Addis, *The Logic of Society* (Minneapolis: University of Minnesota Press, 1975), chapter 11. The point remains, however, that Marx's scientific claims have received insufficient attention.

3. George Brenkert, *Marx's Ethics of Freedom* (London: Routledge & Kegan Paul, 1983); Allen Wood, *Karl Marx* (London: Routledge & Kegan Paul, 1981); Allen Buchanan, *Marx and Justice* (Totowa, N.J.: Littlefield, Adams, 1982); D.-H. Ruben, *Marxism and Materialism*, 2nd ed. (Atlantic Highlands, N.J.: Humanities, 1979). Important recent collections include Terence Ball and James Farr, eds., *After Marx* (Cambridge: Cambridge University Press, 1984); John Mepham and D.-H. Ruben, eds., *Issues in Marxist Philosophy* (Atlantic Highlands, N.J.: Humanities, 1979); and G. H. R. Parkinson, ed., *Marx and Marxisms* (Cambridge: Cambridge University Press, 1982.

4. G. A. Cohen, *Karl Marx's Theory of History: A Defence* (Princeton: Princeton University Press, 1978). John McMurtry, *The Structure of Marx's World-View* (Princeton: Princeton University Press, 1977). William Shaw, *Marx's Theory of History* (Stanford, Calif.: Stanford University Press, 1978). There has also been an active journal literature in this area; see citations in the Bibliography of articles by Jon Elster, John Roemer, G. A. Cohen, Philippe Van Parijs, Gary Young, Richard Miller, and Stanley Moore.

5. Consider, for example, the extensive debate that has developed over G. A. Cohen's claim that historical materialism rests on functionalist explanations. Important contributions include Cohen, "Functional Explanation, Consequence Explanation, and Marxism," *Inquiry* 25 (1982):27–56; Jon Elster, "Marxism, Functionalism, and Game Theory," *Theory and Society* 11 (1982):453–82; Cohen, "Reply to Elster," *Theory and Society* 11 (1982):483–95, John Roemer, "Methodological Individualism and Deductive Marxism," *Theory and Society* 11 (1982):513–20; and Philippe Van Parijs, "Functionalist Marxism Rehabilitated," *Theory and Society* 11 (1982):497–511.

6. John Roemer, *A General Theory of Exploitation* (Cambridge, Mass.: Harvard University Press, 1982) and *Analytic Foundations of Marxian Economic Theory* (Cambridge: Cambridge

University Press, 1981). See also Roemer's many recent journal articles (cited in the Bibliography) on exploitation and related issues.

7. Jon Elster, *Logic and Society* (Chichester: Wiley, 1978) and *Ulysses and the Sirens* (Cambridge: Cambridge University Press, 1979). Related articles are cited in the Bibliography.

8. Robert Brenner, "Agrarian Class Structure and Economic Development in Pre-Industrial Europe," *Past and Present* 70 (1976):30–75, and "The Agrarian Roots of European Capitalism," *Past and Present* 97 (1982):16–113.

9. Erik Olin Wright, *Class, Crisis and the State* (London: Verso, 1978).

10. For a fairly representative sampling of some of the work in this analytic approach to Marxian science, see Terence Ball and James Farr, eds., *After Marx*.

11. Michio Morishima, *Marx's Economics* (Cambridge: Cambridge University Press, 1973); Ian Steedman, *Marx after Sraffa* (London: New Left Books, 1977); Robert Paul Wolff, *Understanding Marx* (Princeton: Princeton University Press, 1984). John Roemer's work is cited in n. 6 above.

12. Althusser and Balibar, *Reading Capital*; Louis Althusser, *For Marx*, trans. Ben Brewster (London: New Left Books, 1969); Nicos Poulantzas, *Political Power and Social Class* (London: New Left Books, 1976); Maurice Godelier, *Rationality and Irrationality in Economics*, trans. Brian Pearce (New York: Monthly Review, 1978); Barry Hindess and Paul Hirst, *Pre-Capitalist Modes of Production* (London: Routledge & Kegan Paul, 1975); Antony Cutler et al., *Marx's 'Capital' and Capitalism Today*, vols. 1, 2 (London: Routledge & Kegan Paul, 1977, 1978).

13. Significant methodological discussions are contained in the introduction to the *Grundrisse*, the first preface to *Capital*, *The Poverty of Philosophy*, *Theories of Surplus Value*, and *Notes on Adolph Wagner*.

14. "The search for a body of rules competent to constitute a given normal research tradition becomes a source of continual and deep frustration. . . . Scientists can agree that a Newton, Lavoisier, Maxwell, or Einstein has produced an apparently permanent solution to a group of outstanding problems and still disagree, sometimes without being aware of it, about the particular abstract characteristics that make those solutions permanent. They can, that is, agree in their identification of a paradigm without agreeing on, or even attempting to produce, a full interpretation or rationalization of it." *The Structure of Scientific Revolutions*, 2nd ed. (Chicago: University of Chicago Press, 1970), p. 44.

15. See Noam Chomsky's discussion of the performance-competence distinction in *Aspects of the Theory of Syntax* (Cambridge, Mass.: MIT Press, 1965), chapter 1. Israel Scheffler applies this distinction to the philosophy of science in *The Anatomy of Inquiry* (Indianapolis: Bobbs-Merrill, 1963), pp. 7–14.

16. Isaac Newton provides a celebrated example of this possibility; his explicit methodological views were highly inductivist (taking expression, for example, in his slogan *hypotheses non fingo*), whereas his creative scientific work gave extensive scope to theory construction. E. A. Burtt, *The Metaphysical Foundations of Modern Science* (New York: Anchor, 1954), pp. 215–26, 239–55. See also Larry Laudan's discussion of the impact of Newton's inductivist theory of science in *Science and Hypothesis* (Dordrecht: Reidel, 1981), pp. 86–103.

Chapter 1. Naturalism and *Capital*

1. V. I. Lenin, *Marx, Engels, Marxism* (Peking: Foreign Languages Press, 1978), p. 15.

2. Ibid., *Marx, Engels, Marxism*, pp. 15, 33.

3. Karl Popper, *The Open Society and Its Enemies*, vol. 2 (Princeton: Princeton University Press, 1966), pp. 84–86.

4. Russell Keat and John Urry, *Social Theory as Science* (London: Routledge & Kegan Paul, 1975), p. 97.

5. Derek Sayer, *Marx's Method* (Atlantic Highlands, N.J.: Humanities, 1979), p. 140.

6. It must be emphasized that the version of naturalism under consideration here is imperfect even in application to the natural sciences. For example, many areas of natural science lack the theoretical unity postulated by the naturalistic model; natural history and geology provide contrary examples. Likewise, many areas of natural science do not describe deterministic laws of nature, e.g., thermodynamics, meteorology, or fluid dynamics. These exceptions do not affect the basic point, however, since we are concerned only with an idealized model of scientific knowledge that many have attempted to apply to the social sciences and to Marx's system; its suitedness or unsuitedness to the natural sciences is irrelevant in this context. All that is important for our purposes is to establish that this model is not well suited to Marx's system.

7. David Thomas, *Naturalism and Social Science* (Cambridge: Cambridge University Press, 1979), p. 1.

8. Roy Bhaskar, "On the Possibility of Social Scientific Knowledge and the Limits of Naturalism," in John Mepham and D.-H. Ruben, eds., *Issues in Marxist Philosophy*, vol. 3 (Atlantic Highlands, N.J.: Humanities, 1979), pp. 107, 108.

9. For an account of some of the varieties of empirical control on theory available to scientific research, see Ernest Nagel, *The Structure of Science* (New York: Harcourt, Brace & World, 1961), pp. 450-59.

10. Keat and Urry, *Social Theory as Science*, p. 25.

11. Ibid., pp. 25, 68, 96; see also Bhaskar, "Possibility of Social Scientific Knowledge," pp. 108-39.

12. It should be noted that the features of science identified here as naturalistic represent a reconstruction of a common theory of science; no single representative of this view endorses all these conditions equally. The general point is that these conditions characterize a large family of theories of science, some of which are in fundamental opposition to others. They represent assumptions that tend to be clustered in many competing views of the nature of scientific knowledge.

13. I use the term "enlightened positivism" to distinguish between the positivism of the Vienna Circle and that of Hempel and his generation. The chief feature concerns the conception of theory at work in the two periods. The early Carnap believed that it would be possible to dispense with a theoretical vocabulary in principle, whereas Hempel and others concluded that theoretical terms were ineliminable from scientific knowledge. For a useful history of the development of positivist theory of science in this century, see Frederick Suppe's historical introduction to Suppe, ed., *The Structure of Scientific Theories*, 2nd ed. (Urbana: University of Illinois Press, 1977).

14. Carl Hempel, "The Theoretician's Dilemma," in Hempel, ed., *Aspects of Scientific Explanation* (New York: Free Press, 1965), p. 183.

15. Ernest Nagel, "The Meaning of Reduction in the Natural Sciences," in Arthur Danto and Sidney Morgenbesser, eds., *Philosophy of Science* (New York: World Publishing, 1960), p. 288.

16. Carl Hempel, "Aspects of Scientific Explanation," in Hempel, ed., *Aspects of Scientific Explanation*, p. 345.

17. Hempel, "The Theoretician's Dilemma," p. 177.

18. Hempel presents a clear exposition and discussion of this view of confirmation; ibid., especially pp. 173-85.

19. Richard Westfall's *The Construction of Modern Science: Mechanisms and Mechanics* (New York: Wiley, 1971) provides an extensive and useful description of this process of development in modern science.

20. See Suppe's account in his extensive introduction to *The Structure of Scientific Theories*, pp. 62-190.

21. Keat and Urry, *Social Theory as Science*, chapter 5; Roy Bhaskar, "Possibility of Social Scientific Knowledge." pp. 109-10; Derek Sayer, *Marx's Method*, pp. 114-15.

22. See, for example, Michio Morishima's *Marx's Economics* (Cambridge: Cambridge University Press, 1973). It should be noted that these arguments are not intended to cast doubt on the value

of Morishima's treatment of Marx's economics, but only to undermine the expectation that it might be possible to assimilate *Capital* to a theoretical construction in which economic premises function as the foundation of a deductive system, and the rest of the book serves as corollary or illustration to these premises.

23. "In *Capital* we are given the elaborated concept of a particular variant form of the general structure of production, namely, the capitalist mode of production. . . . Our proof [and by implication, Marx's] of the validity of the concept of a particular mode of production . . . involves the specification of its relations of production, the deduction of the corresponding forces of production, and the demonstration that the relations and forces form an articulated combination." Barry Hindess and Paul Hirst, *Pre-Capitalist Modes of Production* (London: Routledge & Kegan Paul, 1975), pp. 6, 18. See E. P. Thompson's criticisms of the idealism implicit in this approach to *Capital* in *The Poverty of Theory and Other Essays* (New York: Monthly Review, 1978) (discussed below in chapter 7).

24. See Anthony Brewer, *A Guide to Marx's Capital* (Cambridge: Cambridge University Press, 1984), for a useful commentary on *Capital*.

25. This conclusion follows from the fact that the LTV is generally dispensable in Marx's explanations. Most of the central tendencies that Marx attributes to the CMP are established on the basis of reasoning about rational behavior within the institutions that define the CMP, without making essential use of the concept of value. These "laws of motion" are derived from general facts about the CMP—competition, profits, accumulation, private property, class, and wage labor. And the latter concepts are not dependent on the labor theory of value. This eliminability suggests that the LTV is not part of the empirical hypothesis of *Capital*. See Ian Steedman's *Marx after Sraffa* (London: New Left Books, 1977), pp. 47-49, for arguments supporting a similar conclusion concerning the eliminability of the labor theory of value.

26. This is not to say, however, that philosophers of science universally accept scientific realism; on the contrary, this issue remains highly divisive.

27. Keat and Urry, *Social Theory as Science*, pp. 96-97.

28. Roy Bhaskar, "Possibility of Social Scientific Knowledge," pp. 107-40. See also Edward Nell, "Economics—The Revival of Political Economy," in Robin Blackburn, ed., *Ideology in Social Science* (New York: Vintage, 1973), pp. 94-95.

29. Derek Sayer provides a similar interpretation of the "empiricism" contained in *Notes on Adolph Wagner* in "Science as Critique," in Mepham and Ruben, eds., *Issues in Marxist Philosophy*, vol. 3, pp. 27-54.

30. Ernest Mandel, introduction to *Capital I*, pp. 44-46.

31. This characteristic of social phenomena leads Max Weber to the view that social theory is inherently incomplete: Different research traditions select different "relevant" or "significant" factors out of the infinite variety of social facts and provide causal analyses accordingly. "There is no absolutely 'objective' scientific analysis . . . of 'social phenomena' independent of special and 'one-sided' viewpoints according to which . . . they are selected, analyzed and organized for expository purposes." *The Methodology of the Social Sciences*, trans. and ed. E. Shils and H. A. Finch (New York: Free Press, 1949), p. 72. Weber believes that causal relations can be analyzed objectively, but that there can be no final social theory that orders all causal processes among social phenomena (pp. 79-86).

32. See, for example, E. A. Burtt's description of Galileo's view of natural phenomena: "Nature presents herself to Galileo . . . as a simple, orderly system, whose every proceeding is thoroughly regular and inexorably necessary. . . . Nature is 'inexorable,' acts only 'through immutable laws which she never transgresses.' " *The Metaphysical Foundations of Modern Science* (Garden City, N. Y.: Doubleday, 1954), pp. 74-75. This treatment emphasizes strictly universal laws of nature, that is indeed the preferred form of law within the naturalistic conception. (This is what lies behind Einstein's celebrated aphorism that "God does not play dice.")

33. See chapters 5 and 7 for a more complete discussion of Marx's derivation of this law.

34. See Paul Sweezy, *The Theory of Capitalist Development* (New York: Monthly Review, 1968), pp. 97ff.

35. D.-H. Ruben provides a roughly similar analysis of the logic of laws of tendency in Parkinson, ed., *Marx and Marxisms* (Cambridge: Cambridge University Press, 1982), "Marx, Necessity and Science," pp. 50-56. He further holds, however, that laws of tendency are in principle eliminable from Marx's system, to be replaced by fully specific statements of sufficient conditions (p. 51). On the view taken here, however, according to which laws of tendency depend on the choices made by large numbers of independent agents, laws of tendency are permanently indeterminate. But see also the alternative treatment provided by Roy Bhaskar in "Possibility of Social Scientific Knowledge," pp. 133-34.

36. Mackie discusses this problem in terms of the distinction between a cause and a set of standing conditions. "Causal statements are commonly made in some context, against a background which includes the assumption of some causal field." Mackie, *The Cement of the Universe* (Oxford: Oxford University Press, 1974), pp. 34-35. That is, in order to make a causal ascription at all, we must assume that the appropriate background conditions remain fixed.

37. See n. 31 above.

38. Herbert Simon emphasizes the differences between natural and social science along lines roughly similar to these in *The Sciences of the Artificial*, 2nd ed. (Cambridge, Mass.: MIT Press, 1981).

39. For a brief statement of Marx's view of history as a series of property forms, see *The German Ideology*, pp. 43-46, 68-79. Central to this account is the idea that historical epochs are chiefly distinguished by the social relations through which the process of production is controlled and its fruits distributed: communal ownership (tribal communism), ownership of the direct producer (slavery), ownership of rights to labor services (feudal property), and ownership of capital and labor power (bourgeois property).

40. For an extensive description of the institutional context of peasant agriculture in Southeast Asia, see James Scott, *The Moral Economy of the Peasant* (New Haven: Yale University Press, 1976).

41. Perry Anderson's *Passages from Antiquity to Feudalism* (London: New Left Books, 1974) and *Lineages of the Absolutist State* (London: New Left Books, 1974) provide strong examples of Marxist efforts to give historical answers to questions of these sorts.

42. Robert Brenner offers an extensive development of just this sort of argument in "The Agrarian Roots of European Capitalism," *Past and Present* 97 (1982):16-113. Brenner maintains that the particular circumstances of class and property in late feudalism, and the opportunities and barriers those circumstances imposed on the relevant agents, gave rise to important long-term structural tendencies, e.g., the relative stagnation of feudal technology.

43. Thomas Schelling's *Micromotives and Macrobehavior* (New York: Norton, 1978) develops several examples of such institutional logics, e.g., patterns of residential segregation that emerge from relatively tolerant minimal requirements of neighborhood mix on the part of two distinct subpopulations (chapter 4).

44. Even on this account the naturalistic model creates the danger that the "logic" such analysis uncovers may be hypostatized into an occult power in analogy with laws of nature. We may avoid this risk by recalling that the regularities to which an institutional logic gives rise are always created through individual action, not the exceptionless workings of natural processes.

45. Robert Brenner, "The Agrarian Roots of European Capitalism," p. 16.

46. Jon Elster, "Marxism, Functionalism, and Game Theory,"*Theory and Society* 11 (1982):453.

47. John Roemer, *A General Theory of Exploitation* (Cambridge, Mass.: Harvard University Press, 1982). See also "New Directions in the Marxian Theory of Exploitation," *Politics & Society* 11:3 (1982):253-87, for a summary discussion of Roemer's main ideas on exploitation.

48. It should be noted that these approaches are at odds with the functionalist approach to Marxism argued by G. A. Cohen because they reject the teleology implicit in the functionalist theory. See chapter 5 for a more extensive discussion of this issue.

49. This account makes Marx's explanations rather similar to many other examples from social science, for example, Weber's account of social action, Durkheim's account of mass phenomena such as rates of suicide, and even marginalist economic accounts of price phenomena. Each rests on an analysis of the consequences for the social system of the structure of incentives within which participants make choices and act. The chief difference between Marx and other social scientists concerns the assumptions each makes about the social institutions that most influence individual action.

50. Ralph Miliband, *The State in Capitalist Society* (New York: Basic, 1969).

51. Samuel Bowles and Herbert Gintis, *Schooling in Capitalist America* (New York: Basic, 1976).

52. Harry Braverman, *Labor and Monopoly Capital* (New York: Monthly Review, 1974).

53. For treatments of Marxism as a program of research, see Terrell Carver, *Marx's Social Theory* (Oxford: Oxford University Press, 1982), pp. 22-24, and Allen Wood, *Karl Marx* (London: Routledge & Kegan Paul, 1981), p. 102. In chapter 2 it will be argued that historical materialism is best construed as a research program. Note the parallel between this account of current Marxist social science and Lukács's classical formulation: "Orthodox Marxism, therefore, does not imply the uncritical acceptance of the results of Marx's investigations. . . . On the contrary, orthodoxy refers exclusively to *method*." Lukács, "What Is Orthodox Marxism?" in his *History and Class Consciousness* (Cambridge, Mass.: MIT Press, 1971), p. 1.

Chapter 2. Historical Materialism and *Capital*

1. Particularly important are G. A. Cohen, *Karl Marx's Theory of History: A Defence* (Princeton: Princeton University Press, 1978); John McMurtry, *The Structure of Marx's World-View* (Princeton: Princeton University Press, 1977); William Shaw, *Marx's Theory of History* (Stanford, Calif.: Stanford University Press, 1978); and Melvin Rader, *Marx's Interpretation of History* (Oxford: Oxford University Press, 1979).

2. Cohen's *Karl Marx's Theory of History* provides a thorough reconstruction of these theses.

3. This thesis is applied to the destruction of feudal society by emerging capitalist activity in *The Communist Manifesto* (R1848, pp. 72-79) and in *The German Ideology*: "In the development of productive forces there comes a stage when productive forces and means of intercourse are brought into being, which, under the existing relationships, only cause mischief, and are no longer productive but destructive forces" (*GI*, p. 94).

4. For detailed discussion of the economic and social institutions that organized peasant agriculture within the precolonial Vietnamese village system, see Samuel Popkin, *The Rational Peasant* (Berkeley: University of California Press, 1979), chapters 2 and 3. Popkin holds that the incentives and constraints that defined the positions of peasants, landlords, and "notables" determined the fundamental dynamic of development of the village system.

5. Cohen distinguishes between social and material properties of society. Social properties "entail an ascription to persons of rights or powers vis-a-vis other men"; *Karl Marx's Theory of History*, p. 94. That a given society utilizes a particular technology of steel production is a material fact; that steel factories are owned as *capital* is a social fact. The economic structure consists of *social* relations of production, whereas work relations are material relations of production. Therefore work relations do not belong to the economic structure.

6. That this is the substance of Marx's view is evident in *The German Ideology*. Consider, for example, his account of the sequence of historical stages from tribal communism to capitalism (pp. 43-46). These stages are defined in terms of the property forms that characterize them: tribal ownership, ownership of the slave, landed property with serf labor, private property in money and capital.

7. It will be argued later in this chapter, however, that *Capital* represents only the first step in the full materialist treatment of capitalist society. It does not attempt to explain the superstructural institutions of capitalism, but only to provide the theory of the economic structure on the basis of which such explanations may be constructed.

8. Cohen, *Karl Marx's Theory of History*, pp. 79-85.

9. Richard Miller has an alternative view of the importance of the mode of production in historical materialism. He holds that this concept is in fact the central unit of analysis within the theory of historical materialism. It seems plain, however, that any claim that may be couched in terms of a precisely specified concept of the mode of production on Miller's account may be paraphrased in terms of the economic structure and the forces and relations of production. Moreover, Cohen's arguments concerning the systemic ambiguity and vagueness of the term across Marx's writings seem correct. In any case, the substantive issues that divide Miller and Cohen do not turn on the importance of the notion of the mode of production, and Miller's critique of technological determinism is entirely independent of this issue. Richard Miller, "Productive Forces and the Forces of Change," *The Philosophical Review* 90 (1981):105-12.

10. See Nicos Poulantzas, *Political Power and Social Class*, trans. Timothy O'Hagan (London: New Left Books, 1975), pp. 13-16, for a detailed and relatively clear discussion of the distinction between the mode of production and the social formation.

11. See McMurtry, *Structure of Marx's World-View*, chapter 8, for an extensive treatment of this issue; and Cohen, *Karl Marx's Theory of History*, chapter 6, for Cohen's view of the primacy of the forces of production.

12. Cohen, *Karl Marx's Theory of History*, pp. 152-60.

13. Miller, "Productive Forces," pp. 99-101. See also Joshua Cohen's extensive review of *Karl Marx's Theory of History* (*The Journal of Philosophy* 79 [1982]:253-73), for related arguments against G. A. Cohen's case for the primacy of the forces of production.

14. Cohen, *Karl Marx's Theory of History*, p. 160.

15. McMurtry, *Structure of Marx's World-View*, pp. 189-93.

16. Gary Young, "The Fundamental Contradiction of Capitalist Production," *Philosophy and Public Affairs* 5 (1976):196-234.

17. It is worth noting that Young bases his argument against technological determinism on what he calls Marx's "mature theory" (p. 201) (his economics). Cohen's argument *for* technological determinism, on the other hand, derives chiefly from texts that predate the economics: *The German Ideology* and the preface to *A Contribution to a Critique of Political Economy*. This suggests that Marx's position in *Capital* is significantly different from that in *The German Ideology*: The latter attributes primacy to technology, whereas the former attributes primacy to the social relations of production. If this interpretation is correct, it strengthens the position taken throughout this chapter, namely, that *Capital* and historical materialism are quite independent.

18. One might take this argument to an even stronger conclusion: that the theory of historical materialism, with its emphasis on the centrality of the forces and relations of production in the process of historical change, represents a hasty generalization to all of history from the specific – and unique – characteristics of capitalism. For the bulk of human history has not been characterized by the rapid development of the forces and relations of production found within capitalism; and many social orders (e.g., medieval society) have assigned a more central role for noneconomic institutions (e.g., religion or politico-military institutions) than such institutions play in capitalism. See R. F. Atkinson, "Historical Materialism," in G. H. R. Parkinson, ed., *Marx and Marxisms* (Cambridge: Cambridge University Press, 1982), p. 59.

19. Cohen, *Karl Marx's Theory of History*, pp. 73-77; Perry Anderson, *Arguments Within English Marxism* (London: New Left Books, 1980), pp. 40-43. Both Cohen and Anderson direct their arguments against E. P. Thompson's view that class must be defined in terms of class consciousness; *The Making of the English Working Class* (New York: Vintage, 1963), pp. 9-10.

20. Marx believed that the political awareness of the proletariat would grow out of the objective conditions of life and work that accompanied the objective definition of the proletariat as a class. An important expression of this view occurs in the penultimate chapter of volume I of *Capital*: "Along with the constant decrease in the number of capitalist magnates . . . there also grows the revolt of the working class, a class constantly increasing in numbers, and trained, united and organized by the very mechanism of the capitalist process of production" (*Capital I*, p. 929). Important problems of collective rationality have arisen concerning the issue of "revolutionary motivation," however: the question of the motives rational proletarians might have to contribute to a revolutionary movement. See Allen Buchanan, "Revolutionary Motivation and Rationality," *Philosophy and Public Affairs* 9(1979):59–82, for extensive discussion of this issue (to which we will return in chapter 5 below).

21. Cohen, *Karl Marx's Theory of History*, pp. 148–50.

22. "[Marx] is not saying that noneconomic phenomena are uniquely determined by the economic order. The wide-spread tendency to read 'uniquely determine' when Marx writes 'determine' . . . is one of the great banes of critical commentary on Marx." John McMurtry, *Structure of Marx's World-View*, p. 158. McMurtry argues rather for a form of determination based on attrition: Some noneconomic forms tend to extinguish more readily than do others, given the features of the economic structure.

23. McMurtry, *Structure of Marx's World-View*, p. 161. See also Terrell Carver, *Marx's Social Theory* (Oxford: Oxford University Press, 1982), pp. 38–39, for a similar view.

24. This is the historically conditioned consciousness that Marx refers to in *The Eighteenth Brumaire*: "Men make their own history, but not of their own free will; not under circumstances they themselves have chosen but under the given and inherited circumstances with which they are directly confronted" (*SE*, p. 146).

25. Cohen, *Karl Marx's Theory of History*, p. 160.

26. Ibid.

27. Cohen, "Functional Explanation, Consequence Explanation, and Marxism," *Inquiry* 25 (1982):27.

28. Cohen, *Karl Marx's Theory of History*, pp. 285–89.

29. "We can rationally hypothesize functional explanations even when we lack an account which, like Darwin's, shows how the explanations work, or . . . even when we lack elaborations of the explanations." Cohen, *Karl Marx's Theory of History*, pp. 285–86.

30. Jon Elster, "Marxism, Functionalism, and Game Theory," *Theory and Society* 11 (1982):453.

31. Ibid., p. 463.

32. Ibid., p. 454.

33. John Roemer, "Methodological Individualism and Deductive Marxism," *Theory and Society* 11 (1982):513.

34. Antonio Gramsci, *Selections from the Prison Notebooks of Antonio Gramsci*, trans. and ed. Q. Hoare and G. Nowell-Smith (New York: International Publishers, 1971).

35. Ralph Miliband, *The State in Capitalist Society* (New York: Basic, 1969).

36. Poulantzas, *Political Power and Social Classes*, p. 12.

37. I. I. Rubin, *Essays on Marx's Theory of Value*, trans. Milos Amardzija and Fredy Perlman (Detroit: Black and Red, 1972; originally published in 1928), pp. 1, 2.

38. Maurice Godelier, *Rationality and Irrationality in Economics* (New York: Monthly Review, 1972), p. 198.

39. McMurtry, *Structure of Marx's World-View*, p. 91.

40. Cohen, *Karl Marx's Theory of History*, p. 27.

41. For example, chapter 7 of *Karl Marx's Theory of History* is an extensive account of the capi-

talist mode of production in Marx's theory of history that draws very heavily on citations from *Capital* and *Theories of Surplus Value*.

42. Thompson espouses a position in *The Poverty of Theory and Other Essays* (New York: Monthly Review, 1978) that is diametrically opposed both to the position argued here and to the subsumption theory as well. This may be called the deviation theory: Marx's economic research was an untoward distraction from his more fruitful work in historical materialism (pp. 59, 65, 164).

43. "My inquiry led me to the conclusion that neither legal relations nor political forms could be comprehended . . . by themselves, . . . but that on the contrary they originate in the material conditions of life; . . . that the anatomy of this civil society, however, has to be sought in political economy" (*EW*, p. 425).

44. Consider, for example, Marx's critique of Proudhon's political economy: "M. Proudhon the economist understands very well that men make cloth, linen or silk materials in definite relations of production. But what he has not understood is that these definite social relations are just as much produced by men as linen, flax, etc. Social relations are closely bound up with productive forces" (*PP*, p. 109).

45. The chief writings of historical materialism include at least *The German Ideology*, *The Communist Manifesto*, and the preface to *A Contribution to A Critique of Political Economy*.

46. This formulation depends chiefly on the preface to *A Contribution to A Critique of Political Economy*.

47. Roman Rosdolsky's *The Making of Marx's Capital* (London: Pluto Press, 1977) offers a detailed discussion of the evolution of Marx's plans of research (pp. 10-62). One such plan is found in the Preface to the *Grundrisse* (p. 108).

48. An important exception to this remark is Marx's historical analysis of the political conflicts in France; these writings are detailed works of historical investigation. See *The Class Struggles in France*, *The Eighteenth Brumaire of Louis Napoleon*, and *The Civil War in France* (*SE*, pp. 35-142; 143-249; *FI*, pp. 187-236).

49. *The German Ideology*, pp. 43-46. Perry Anderson's *Passages from Antiquity to Feudalism* (London: New Left Books, 1974) covers much the same terrain, but in substantially greater detail.

50. E. J. Hobsbawm arrives at a rather similar assessment of Marx and Engels's use of evidence in his introduction to *Pre-Capitalist Economic Formations* (New York: International Publishers, 1965): "It is generally agreed that Marx and Engels' observations on pre-capitalist epochs rest on far less thorough study than Marx's description and analysis of capitalism. Marx concentrated his energies on the study of capitalism, and he dealt with the rest of history in varying degrees of detail, but mainly in so far as it bore on the origins and development of capitalism" (p. 20).

51. The importance of these materials to Marx is apparent from his correspondence: "One Blue Book after another arrived while I was in the midst of the final elaboration, and I was delighted to find my theoretical results fully confirmed by the facts" (*Correspondence*, pp. 180-81).

52. See, for example, *Capital I*, p. 802.

53. See *Capital II*, pp. 109-229, and *Capital III*, pp. 200-34, 273-301, 317-75.

54. Thomas Kuhn, *The Structure of Scientific Revolutions*, 2nd ed. (Chicago: University of Chicago Press, 1970); Imre Lakatos, "The Methodology of Scientific Research Programmes," in Imre Lakatos and Alan Musgrave, eds., *Criticism and the Growth of Knowledge* (Cambridge: Cambridge University Press, 1970). Larry Laudan's *Progress and Its Problems* (Berkeley: University of California Press, 1977) and Harold Brown's *Perception, Commitment, and Theory* (Chicago: University of Chicago Press, 1979) provide useful summaries of recent treatments of the framework of assumptions that guide scientific research.

55. Terrell Carver offers a similar view of the significance of historical materialism. In discussing the 1859 preface he writes, "I take it that by general result and guiding thread what Marx had in mind was more like a hypothesis for research than a law on the model of the physical sciences

or the social sciences (whether conceived deterministically or not)." *Marx's Social Theory*, pp. 22-23. See also Allen Wood's *Karl Marx*, p. 102.

56. Maurice Godelier provides an extensive discussion of *Capital* as a work of social science in *Rationality and Irrationality in Economics*, pp. 130-61.

Chapter 3. Marx's Economic Analysis

1. The labor theory of value remains a central issue in Marxist economic theory. Important contributions to the literature include I. I. Rubin, *Essays on Marx's Theory of Value*, trans. Milos Amardzija and Fredy Perlman (Detroit: Black and Red, 1972; originally published in 1928); Paul Sweezy, *The Theory of Capitalist Development* (New York: Monthly Review, 1968); Michio Morishima, *Marx's Economics* (Cambridge: Cambridge University Press, 1973); Joan Robinson, *An Essay in Marxian Economics* (New York: St. Martin's, 1942); David Horowitz, ed., *Marx and Modern Economics* (New York: Modern Reader, 1968); and E. K. Hunt and Jesse G. Schwartz, eds., *A Critique of Economic Theory* (Baltimore: Penguin, 1972).

2. See John Roemer's *A General Theory of Exploitation and Class* (Cambridge, Mass.: Harvard University Press, 1982) for a general treatment of the inequalities inherent in differential ownership of property. Roemer holds that exploitation is not specific to the wage relation, but rather is a general feature of any property system in which a minority has exclusive control over the productive forces.

3. Ian Steedman, *Marx after Sraffa* (London: New Left Books, 1977), chapter 2.

4. Morishima, *Marx's Economics*, chapter 9.

5. Steedman, *Marx after Sraffa*, chapter 2.

6. Marx's position is in common with other classical political economists. Competition, or the relation of supply to demand, can explain fluctuations of market price around an equilibrium price, but cannot explain the point of equilibrium itself. If market conditions result in a prolonged difference between the value of a commodity and its market price, investors will enter or exit the market, resulting in increases or reductions in the supply of the good. The result is that the price will return to its value. Thus deviation of price from value leads—through competition—to the restoration of the equilibrium price, determined by the real cost of producing the good. The fundamental determinant, therefore, is the process of production, which determines the value of the good and its cost of production. See Ronald Meek's discussion of Marx's view of demand in *Studies in the Labor Theory of Value*, 2nd ed. (New York: Monthly Review, 1956), pp. 177-78.

7. Marx's own treatment of the transformation problem was mathematically flawed. More satisfactory treatments are provided by Shigeto Tsuru in "On Reproduction Schemes," in Paul Sweezy's *The Theory of Capitalist Development* (New York: Monthly Review, 1968; appendix A), and by Morishima, *Marx's Economics*, chapter 7.

8. Note, however, John Roemer's argument that labor value depends on the prices of commodities because it is not possible to determine socially necessary labor time independently from equilibrium price. "New Directions in the Marxian Theory of Exploitation," *Politics and Society* 11 (1982):270-73.

9. Allen Wood, *Karl Marx* (London: Routledge & Kegan Paul, 1981), p. 225.

10. Piero Sraffa, *The Production of Commodities by Means of Commodities* (Cambridge: Cambridge University Press, 1960). As will be briefly discussed in the next section, Sraffa's work represents a modern analysis of the capitalist economy that is rooted in classical political economy. Sraffa's work is seen by some Marxist economists as offering an alternative to the labor theory of value. He examines the production process as a set of physical inputs and outputs.

11. Morishima, *Marx's Economics*, pp. 10-12.

12. "The modern theory of prices does not owe anything to the Marxian version of the classical labour theory of value nor can it in my opinion profit from any attempts towards reconciliation or mediation between the two types of approach." Wassily Leontief, "The Significance of Marxian Eco-

nomics for Present-day Economic Theory," in David Horowitz, ed., *Marx and Modern Economics* (New York: Modern Reader, 1968), p. 88.

13. Böhm-Bawerk argues against the labor theory of value on the grounds that it is justified on a "dialectical" rather than an empirical basis; it is therefore inaccessible to the methods of empirical reasoning. Eugen von Böhm-Bawerk, *Karl Marx and the Close of His System* (London: Merlin, 1975), pp. 64-70. (We will offer a similar argument in chapter 6, but without drawing the destructive conclusions Böhm-Bawerk emphasizes.)

14. See Ronald Meek, *Studies in the Labor Theory of Value*, chapter 6, for a survey of criticisms of the labor theory of value.

15. Robinson, *An Essay on Marxian Economics*, p. 22.

16. Ian Steedman, *Marx after Sraffa*, chapters 3-5.

17. Michio Morishima, *Marx's Economics*, pp. 1-2.

18. Ibid., p. 3.

19. Ibid., pp. 3-4.

20. Ibid., chapter 5.

21. Ibid., p. 4.

22. Ibid., p. 194.

23. See also Steedman, ed., *The Value Controversy* (London: New Left Books, 1981), for a useful collection of responses to Steedman's main claims.

24. *Marx after Sraffa*, p. 48.

25. Steedman, *The Value Controversy*, pp. 11-19.

26. Ibid., p. 19.

27. Ibid.

28. Max Weber, *The Protestant Ethic and the Spirit of Capitalism* (New York: Scribner's, 1958), p. 181.

29. Steedman, *Marx after Sraffa*, chapter 9.

30. For other arguments to a similar conclusion, see G. A. Cohen, "The Labor Theory of Value and the Concept of Exploitation," *Philosophy and Public Affairs* 8 (1979):338-60, and Wood, *Karl Marx*, pp. 231-34. See also Nancy Holmstrom, "Marx and Cohen on Exploitation and the Labor Theory of Value," *Inquiry* 26 (1983):287-307, for a rebuttal of Cohen's position.

31. Roemer, "New Directions in the Marxian Theory of Exploitation," pp. 274-75.

32. Ibid., p. 268.

33. John Roemer, "Property Relations vs. Surplus Value in Marxian Exploitation," *Philosophy and Public Affairs* 11 (1982):281.

34. Ibid., p. 284.

35. John Roemer, "R. P. Wolff's Reinterpretation of Marx's Labor Theory of Value," *Philosophy and Public Affairs* 10 (1981):81.

Chapter 4. Essentialism, Abstraction, and Dialectics

1. Consider, for example, the distance between Isaac Newton's explicit theory of science and his actual scientific practice (discussed in n. 16 in the introduction).

2. As Norm Geras writes, "Marx . . . provides us with a conception of the minimum necessary condition to be satisfied by any work aspiring to scientific status: namely, that it uncover the reality behind the appearance which conceals it." "Marx and the Critique of Political Economy," in Robin Blackburn, ed., *Ideology in Social Science* (New York: Vintage, 1973), p. 285.

3. For statements of Marx's essentialism, see Georg Lukács, *History and Class Consciousness* (Cambridge, Mass.: MIT Press, 1971), pp. 7-8; Maurice Godelier, *Rationality and Irrationality in Economics*, trans. Brian Pearce (New York: Monthly Review, 1973), p. 136; and Louis Althusser and Etienne Balibar, *Reading Capital*, trans. Ben Brewster (London: New Left Books, 1970), p. 83.

4. G. A. Cohen, *Karl Marx's Theory of History: A Defence* (Princeton: Princeton University Press, 1978), p. 326.

5. Roman Rosdolsky, *The Making of Marx's Capital* (London: Pluto Press, 1977), p. 51.

6. Althusser and Balibar, *Reading Capital*, p. 83.

7. Marx distinguishes between classical political economy and vulgar political economy in the following terms: "Let me point out once and for all that by classical political economy I mean all the economists who, since the time of W. Petty, have investigated the real internal framework of bourgeois relations of production, as opposed to the vulgar economists who only flounder around within the apparent framework of those relations, ceaselessly ruminate on the materials long since provided by scientific political economy, and seek there plausible explanations of the crudest phenomena for the domestic purposes of the bourgeoisie" (*Capital II*, p. 175). Thus the distinction between classical and vulgar political economy is exactly parallel to the fundamental distinction represented by essentialism: Classical political economy seeks to penetrate to the underlying social relations, whereas vulgar political economy is content to remain at the level of misleading appearances.

8. "The Fetishism of Commodities" (*Capital II*, pp. 163–77) may be interpreted as a thesis in the philosophy of social science that insists on the need for abstract hypotheses in a theory of capitalism. Visible social relations – market relations among commodity producers – are mystifying because they obscure the underlying relations of production.

9. See, for example, Noam Chomsky's review of B. F. Skinner's *Verbal Behavior* (*Language* 35 [1959]:26–58). Significantly, Chomsky's own theory of science is essentialist in many of the same ways that Marx's is. Thus Chomsky insists on the reality of unobservable linguistic structures and holds forcefully that explanation in linguistics can proceed only on the basis of strong theories of unobservable mental processes.

10. G. A. Cohen, "Karl Marx and the Withering Away of Social Science" in his *Karl Marx's Theory of History* (appendix), provides a useful discussion of this point.

11. "The vulgar economist does nothing more than translate the peculiar notions of the competition-enslaved capitalists into an ostensibly more theoretical and generalized language, and attempt to demonstrate the validity of these notions" (*Capital III*, p. 338).

12. "One corollary of the dictum on reality and appearance is that science may study a social formation only if it is held together by mechanisms that disguise its basic anatomy." Cohen, *Karl Marx's Theory of History*, p. 336.

13. Ralph Miliband analyzes some of the mechanisms through which consciousness is shaped so as to make a class regime more tolerable in *Marxism and Politics* (Oxford: Oxford University Press, 1977), pp. 43–64, and *The State in Capitalist Society* (New York: Basic, 1969), pp. 219–64.

14. For discussions of the abstractive method, see Ronald Meek, *Studies in the Labor Theory of Value*, 2nd ed. (New York: Monthly Review, 1956), pp. 299–319; Paul Sweezy, *The Theory of Capitalist Development* (New York: Monthly Review, 1968), pp. 11–20; and Gerald Cohen, "Karl Marx and the Withering Away of Social Science."

15. Roman Rosdolsky discusses the methodological import of Marx's critique of Ricardo in *The Making of Marx's Capital*, pp. 561–68.

16. Bertell Ollman develops this point extensively in his theory of internal relations in *Alienation: Marx's Conception of Man in Capitalist Society* (Cambridge: Cambridge University Press, 1971), pp. 14–16, 27–42, 52–61, and elsewhere. "Capital, labor, value, commodity, etc., are all grasped as relations, containing in themselves, as integral elements of what they are, those parts with which we tend to see them externally tied" (p. 15). See also the sharp criticisms of this view, however, advanced by D.-H. Ruben in *Marxism and Materialism*, 2nd ed. (Atlantic Highlands, N.J.: Humanities, 1979), pp. 63–64, 123–24.

17. Marx does hold that knowledge of the modern capitalist economy sheds some light on earlier systems of production, including classical slavery and feudalism, but these insights cannot be

mechanically extracted from the theoretical description of capitalism. "Bourgeois society is the most developed and the most complex historic organization of production. The categories which express its relations, . . . thereby also allow insights into the structure and the relations of production of all the vanished social formations out of whose ruins and elements it built itself up, whose partly still unconquered remnants are carried along within it. . . . The bourgeois economy thus supplies the key to the ancient, etc. But not at all in the manner of those economists who smudge over all historic differences and see bourgeois relations in all forms of society" (*Grundrisse*, p. 105).

18. See Derek Sayer's discussion of this set of issues in *Marx's Method* (Atlantic Highlands, N.J.: Humanities, 1979), pp. 111-13.

19. In his study of the composition of Marx's economics Rosdolsky concurs in this view that the organization of *Capital* directly reflects the methodological assumptions advanced in the *Grundrisse* introduction discussed earlier because it systematically proceeds from abstract to concrete. "[Marx's] original outline was clearly drawn up so that the process of synthesis, the 'ascent from the abstract to the concrete' occurs there several times. This can be seen particularly clearly in the changes to the outline on page 108 of the *Grundrisse*. In this version, the inquiry proceeds from general categories (exchange value, money, price), through an analysis of the 'inner structure of production'—the categories of capital, landed property and wage-labour—in order to arrive at the synthesis of bourgeois society in the form of the state." "Finally, the same procedure of the ascent from 'the abstract to the concrete' is repeated in the Book on Capital, where Marx begins with 'capital in general' in order to reach, via an examination of competition and the credit system, capital in its most developed form, share-capital." *Making of Marx's Capital*, p. 27.

20. "It is impossible to understand completely Marx's *Capital*, and especially its first chapter, without having thoroughly studied and understood the *whole* of Hegel's *Logic*." V. I. Lenin, *Philosophical Notebooks* (posthumous) in Robert Tucker, ed., *The Lenin Anthology* (New York: Norton, 1975), p. 639.

21. "By adopting the progressive part of Hegel's method, namely the dialectic, Marx not only cut himself off from Hegel's successors; he also split Hegel's philosophy in half." Georg Lukács, "What Is Orthodox Marxism?" *History and Class Consciousness*, p. 17.

22. "The purpose of *Capital* is itself a clear reminder of the method of knowledge applied by Marx to his main work: the method of the materialist dialectic." Ernest Mandel, Introduction, *Capital II*, p. 17.

23. Lenin, *On the Question of Dialectics* (Moscow: Progress Publishers, 1980), pp. 11-12.

24. "For once capital has emerged on the page, its self-development is determined by the innate logic inherent within the category." E. P. Thompson, *The Poverty of Theory and Other Essays* (New York: Monthly Review, 1978), p. 61.

25. "The logical method adopted by Marx is intrinsically and organically the Hegelian one itself; he accepted and applied the Hegelian dialectic . . . , and thus compromised the firmness and efficacy of his system." Benedetto Croce, *Essays on Marx and Russia* (New York: Frederick Ungar, 1966), p. 22.

26. Sidney Hook, *From Hegel to Marx* (Ann Arbor: University of Michigan Press, 1962), pp. 60-75.

27. "In *Capital* we are given the elaborated concept of a particular variant form of the general structure of production, namely, the capitalist mode of production. . . . Our proof [and by implication, Marx's] of the validity of the concept of a particular mode of production . . . involves the specification of its relations of production, the deduction of the corresponding forces of production, and the demonstration that the relations and forces form an articulated combination." Barry Hindess and Paul Hirst, *Pre-Capitalist Modes of Production* (London: Routledge & Kegan Paul, 1975), pp. 6, 18.

28. G. W. F. Hegel, *Hegel's Logic. Part One of the Encyclopaedia of the Philosophical Sciences (1830)*, trans. William Wallace (Oxford: Oxford University Press, 1975), p. 8.

29. G. W. F. Hegel, *Lectures on the Philosophy of World History*, trans. H. B. Nisbet (Cambridge: Cambridge University Press, 1975), p. 28.

30. G. W. F. Hegel, *Reason in History*, trans. Robert S. Hartman (Indianapolis: Bobbs Merrill, 1953), pp. 78-79.

31. Hegel, *Lectures*, p. 30.

32. "This book, then, containing as it does the science of the state, is to be nothing other than the endeavor to apprehend and portray the state as something inherently rational. . . . The science of right is a section of philosophy. Consequently its task is to develop the Idea . . . out of the concept, or what is the same thing, to look on at the proper immanent development of the thing itself." G. W. F. Hegel, *The Philosophy of Right* trans. T. M. Knox (Oxford: Oxford University Press, 1969), pp. 11, 14.

33. It must be acknowledged that examples of Hegelian language and dialectical ideas may be found throughout *Capital*. These include Marx's use of the notion of immanent contradictions ("In a crisis, the antithesis between commodities and their value-form, money, is raised to the level of an absolute contradiction." *Capital II*, pp. 235-40; see also *Capital III*, pp. 357-59, 914); his distinction between essence and appearance ("While it is not our intention here to consider the way in which the immanent laws of capitalist production manifest themselves in the external movement of the individual capitals . . . this much is clear: a scientific analysis of competition is possible only if we can grasp the inner nature of capital, just as the apparent motions of the heavenly bodies are intelligible only to someone who is acquainted with their real motions." *Capital II*, p. 433; see also *Capital III*, pp. 268-69, 338); his notion that the concept of capital has internal logical structure, and is internally related to the concept of labor power (*Capital III*, pp. 139; 953 -57); his distinction between quantitative and qualitative change (*Capital II*, pp. 439-54); very infrequently, his use of the term "dialectics" itself ("To the extent that commodity production, in accordance with its own immanent laws, undergoes a further development into capitalist production, the property laws of commodity production must undergo a dialectical inversion so that they become laws of capitalist appropriation." *Capital II*, pp. 733-34); and in what is essentially a coda to the first volume of *Capital* ("The Historical Tendency of Capitalist Accumulation"), Marx's description of the laws of capitalist accumulation in terms of the negation of the negation (*Capital II*, p. 929). One might also consider the dialectical form in which Marx casts the relations between production, distribution, exchange, and consumption in the *Grundrisse* pp. 88-100). Significantly, however, this treatment is abandoned in *Capital* in favor of an analysis that is based on the historical and institutional characteristics of these aspects of the capitalist mode of production (*Capital II*, pp. 273-76; *Capital III*, pp. 1017-24). What is striking about these examples, however, is not that they occur, but that they occur so infrequently in *Capital*, and that they play no role whatsoever in substantive explanations or analysis in that work.

34. This description of Marx's method appears to be consistent with Mandel's endorsement of dialectics. Mandel, Introduction, *Capital II*, pp. 17-19.

35. As Allen Wood puts it, "[Marx] accepts Hegel's vision of reality but rejects the Hegelian metaphysical underpinnings of this vision, together with the epistemological conclusions that are supposed to follow from them. For Marx the world is a system of organically interconnected processes characterized by inherent tendencies to development, and subject periodically to radical changes in organic structure." Wood, *Karl Marx* (London: Routledge & Kegan Paul, 1981), p. 208.

36. Hegel, *Reason in History*, pp. 78-79.

37. D.-H. Ruben, "Marxism and Dialectics," in John Mepham and D.-H. Ruben, eds., *Issues in Marxist Philosophy*, vol. 1 (Atlantic Highlands, N.J.: Humanities, 1979), p. 38.

38. Terrell Carver, *Marx's Social Theory* (Oxford: Oxford University Press, 1982), p. 45.

39. Milton Fisk, "Dialectic and Ontology," in John Mepham and D.-H. Ruben, eds., *Issues in Marxist Philosophy*, vol. 1, pp. 117-43.

40. Fisk, "Dialectic and Ontology," pp. 119-23.

41. James Farr, "Marx No Empiricist," *Philosophy of the Social Sciences* 13 (1983):465-72; James Farr, "Marx and Positivism," in Terence Ball and James Farr, eds., *After Marx* (Cambridge: Cambridge University Press, 1984), pp. 217-34. See also Norman Geras, "Marx and the Critique of Political Economy."

42. E. A. Burtt describes Galileo's conception of science in terms strikingly like Marx's: "The world of the senses is not its own explanation; as it stands it is an unsolved cipher, a book written in a strange language, which is to be interpreted or explained in terms of the alphabet of that language." *The Metaphysical Foundations of Modern Science*, 2nd ed. (New York: Anchor, 1954), pp. 78-79.

43. Richard Westfall, *The Construction of Modern Science* (New York: Wiley, 1971), p. 29.

44. Burtt, *Metaphysical Foundations*, p. 79.

45. Ibid., pp. 79-80.

Chapter 5. Explanation

1. Important contributions to this approach to Marxism include John Roemer, *A General Theory of Exploitation and Class* (Cambridge, Mass.: Harvard University Press, 1982) and *Analytic Foundations of Marxian Economic Theory* (Cambridge: Cambridge University Press, 1981); Jon Elster, *Logic and Society* (Chichester: Wiley, 1978) and *Ulysses and the Sirens* (Cambridge: Cambridge University Press, 1979); Robert Brenner, "Agrarian Class Structure and Economic Development in Pre-Industrial Europe," *Past and Present* 70 (1976):30-75, and "The Agrarian Roots of European Capitalism," *Past & Present* 97 (1982):16-113; Philippe Van Parijs, *Evolutionary Explanation in the Social Sciences* (Totowa, N.J.: Rowman & Littlefield, 1981).

2. Ralph Miliband provides this sort of detailed account of the behavior of the capitalist state and the pathways by which state policy comes to reflect capitalist interests throughout his work. See particularly *Capitalist Democracy in Britain* (Oxford: Oxford University Press, 1982), and *The State in Capitalist Society* (New York: Basic, 1969). For an engaging non-Marxist account of some of the diverse factors that shaped American policy toward China in 1949-50, see Nancy Bernkopf Tucker, *Patterns in the Dust: Chinese-American Relations and the Recognition Controversy 1949-50* (New York: Columbia University Press, 1983).

3. Cohen develops a robust defense of functional explanations and consequence explanations in historical materialism in *Karl Marx's Theory of History: A Defence* (Princeton: Princeton University Press, 1978), chapters 9 and 10. In his subsequent article, "Functional Explanation, Consequence Explanation, and Marxism" (*Inquiry* 25 [1982]:27-56), he distinguishes more carefully between functional explanations and consequence explanations (pp. 35-37). An extensive debate has developed over Cohen's claim that historical materialism rests on functionalist explanations. Important contributions include several articles in a special number of *Theory and Society* (11, 1982): John Elster, "Marxism, Functionalism, and Game Theory," pp. 453-82; G. A. Cohen, "Reply to Elster," pp. 483-95; John Roemer, "Methodological Individualism and Deductive Marxism," pp. 513-20; and Philippe Van Parijs, "Functionalist Marxism Rehabilitated," pp. 497-511. See also essays by Philippe Van Parijs and James Noble in Terence Ball and James Farr, eds., *After Marx* (Cambridge: Cambridge University Press, 1984), pp. 88-104 and 105-19, respectively.

4. Cohen, *Karl Marx's Theory of History*, p. 160.

5. Ibid., p. 226.

6. Consider Albert Szymanski's brief account of the causes of New Deal legislation. After pointing out that the primary business organizations had opposed the New Deal reforms, Szymanski writes: "Only by recognizing that the state bureaucracy has . . . some autonomy from direct control by the majority bloc among the capitalist class, and hence is able to implement policies in the interest of the entire capitalist class, can this be made understandable. Because of the narrowmindedness and internal antagonisms of business in some crisis situations, state bureaucracy must

sometimes develop policies that advance the interests of the system even when the primary beneficiaries of these policies drag their feet." *The Capitalist State and the Politics of Class* (Cambridge, Mass.: Winthrop, 1978), p. 271.

7. In this vein Elster writes that "by assimilating the principles of functionalist sociology . . . Marxist social analysis has acquired an apparently powerful theory that in fact encourages lazy and frictionless thinking." Elster, "Marxism, Functionalism, and Game Theory," p. 453. See also *Ulysses and the Sirens*, pp. 28-35; *Explaining Technical Change* (Cambridge: Cambridge University Press, 1983), chapter 2; and "Cohen on Marx's Theory of History," *Political Studies* 28 (1980):121-28.

8. Noble, "Marxian Functionalism," in Ball and Farr, eds., *After Marx*, p. 108.

9. Cohen refers to this sort of additional explanatory investigation as providing an "elaboration" of a consequence explanation. But Cohen holds that functionalist explanations may stand by themselves without such elaborations and that the search for micromechanisms stands outside the central subject matter of historical materialism. *Karl Marx's Theory of History*, pp. 285-89, and "Functional Explanation. Consequence Explanation, and Marxism." pp. 33-35, 50-54.

10. Elster provides an example of such an account based on an analogy with natural selection in *Explaining Technical Change*, pp. 141-43.

11. Roemer, "Methodological Individualism and Deductive Marxism," p. 516.

12. Mancur Olson, *The Logic of Collective Action: Public Goods and the Theory of Groups* (Cambridge, Mass.: Harvard University Press, 1965).

13. Roemer, "Methodological Individualism and Deductive Marxism," p. 516.

14. Allen Buchanan, "Revolutionary Motivation and Rationality," *Philosophy and Public Affairs* 9 (1979):63.

15. This problem corresponds to the "prisoner's dilemma" in the theory of games. In the prisoners' dilemma two players are faced with a pair of choices: "confess" and "don't confess." The punishments are arranged in such a way that each individual's best strategy—no matter what the other player chooses—is to confess. But if neither confesses, both would be better off. Thus there is a collective good to be gained that cannot be reached through individual rational choice. See R. D. Luce and Howard Raiffa, *Games and Decisions* (New York: Wiley, 1958) and Anatol Rapoport and A. M. Chammah, *Prisoners' Dilemma: A Study in Conflict and Cooperation* (Ann Arbor: University of Michigan Press, 1965) for an exposition of this problem.

16. This is the approach that Buchanan takes ("Revolutionary Motivation and Rationality," pp. 68-76). This approach remains within the rational-action model and attempts to show that in certain cases organizations can make it prudentially rational for their members to cooperate.

17. Thomas Nagel argues for a conception of practical reason according to which altruistic behavior is fully rational: "Just as there are rational requirements on thought, there are rational requirements on action, and altruism is one of them." The *Possibility of Altruism* (Oxford: Oxford University Press, 1970), p. 3. Derek Parfit presents a careful analysis of some of the paradoxes of altruistic and prudential rationality in *Reasons and Persons* (Oxford: Oxford University Press, 1984), part I. A. K. Sen criticizes the maximizing conception of economic rationality and self-interest because it cannot take account of normative commitments that may rationally limit the range of accessible alternatives; "Rational Fools," in Frank Hahn and Martin Hollis, eds., *Philosophy and Economic Theory* (Oxford: Oxford University Press, 1979). And Jon Elster argues for a conception of "conditional altruism" as a possible solution of repeated problems of collective action in "Marxism, Functionalism, and Game Theory," pp. 468-74.

18. Consider James Scott's research on peasant behavior and the circumstances that lead to collective action, resistance, and rebellion in Southeast Asian peasant communities. Scott argues that there is a distinctive "moral economy of subsistence" underlying patterns of peasant resistance and acquiescence and that bonds of solidarity, kinship, and shared moral judgments play a central role in explaining peasant behavior in the face of changes in systems of land tenure, rent, and taxation.

James Scott, *The Moral Economy of the Peasant* (New Haven: Yale University Press, 1976). See also Samuel Popkin, *The Rational Peasant* (Berkeley: University of California Press, 1979), who argues against the assumption of powerful bonds of village solidarity.

19. "On this account, what distinguishes Marxist from non-Marxist social science is not the postulate of methodological individualism, but the level of aggregation deduced as applicable in social analysis." Roemer, "Methodological Individualism and Deductive Marxism," p. 518.

20. Debate over methodological individualism has a long history, extending back at least to Mill's discussion of social science in his *System of Logic*, and forwarded by Karl Popper and J. W. N. Watkins in the 1950s. Especially important is Watkins, "Methodological Individualism and Social Tendencies," in May Brodbeck, ed., *Readings in the Philosophy of the Social Sciences* (New York: Macmillan, 1968), pp. 269–80.

21. Consider, for example, James O'Connor's account of technical change: "In the United States there clearly has been a long-run increase in the volume of complementary and discretionary physical investment. . . . Because of the increased complexity and interdependence of production, monopoly capital needs more and better physical capital. More specifically, the growing interpenetration of monopoly capital and the state is related to the increased pace of economic and social change arising from the advance of technology." *The Fiscal Crisis of the State* (New York: St. Martin's, 1973), p. 103.

22. Robert Brenner's discussion of medieval agrarian relations illustrates just the opposite result in medieval society. He shows that technical innovations were available that could not be implemented because no social group had both the power and the incentive to introduce the innovation. "Agrarian Roots of European Capitalism," pp. 17–19.

23. See Elster's extensive discussion of these problems in *Explaining Technical Change*, chapter 7.

24. See Ernest Mandel's discussion of this point in his Introduction, *Capital II*, pp. 68–73.

25. First, it has been shown that the rate of profit has in fact moved up *and* down between 1900 and 1960. Second, Ian Steedman contends that the argument given here is invalid because it postulates that profits equal surplus value. Steedman argues that this is demonstrably untrue, so that reasoning about R is irrelevant; R does not equal the rate of profit. *Marx after Sraffa* (London: New Left Books, 1977), pp. 50–67.

26. Paul Sweezy appears to take this view. "In Marx's theory, however, the system of production includes both Industrial Employment and the Reserve Army. . . . The fact remains that [the capitalist system] contains within itself a mechanism for regulating the wage level and hence for maintaining profits." Sweezy, *The Theory of Capitalist Development* (New York: Monthly Review, 1968), p. 91.

27. Thomas Schelling, *Micromotives and Macrobehavior* (New York: Norton, 1978), chapter 2.

28. Luce and Raiffa, *Games and Decisions*, and Anatol Rapoport, *Two-Person Game Theory* (Ann Arbor: University of Michigan Press, 1966) provide clear presentations of the main ideas of rational-choice theory and its companion, the theory of games.

29. Robert Brenner's work provides an especially clear example of this sort of "microclass" analysis. See n. 1 above for citations.

30. Mancur Olson, *The Logic of Collective Action* is a classic text on the conflicts between collective and individual interests. Brian Barry gives a clear exposition of Olson's views in *Sociologists, Economists and Democracy* (Chicago: University of Chicago Press, 1978), pp. 23–47. Kenneth Arrow, *Social Choice and Individual Values* (New York: Wiley, 1963), and A. K. Sen, *Collective Choice and Social Welfare* (San Francisco: Holden-Day, 1970), provide rigorous analysis of some of the procedures by which a democratic society may attempt to aggregate individual choices into a collective choice.

31. Thomas Hobbes's argument from assumptions about rational individuals in the state of nature to conclusions about the state of insecurity and war that results is found in *Leviathan*, chapter 13.

32. Lynn White, *Medieval Technology and Social Change* (Oxford: Oxford University Press, 1962), pp. 41–57. See also Marc Bloch's pivotal discussion of the diffusion of the heavy plow in *French Rural History* (Berkeley: University of California Press, 1966).

33. See Marc Bloch's essay on the waterwheel in *Land and Work in Medieval Europe* (London: Routledge & Kegan Paul, 1967). Terry Reynolds provides a survey of the extensive use of water power in medieval manufacture in "Medieval Roots of the Industrial Revolution," *Scientific American* 251 (1984):122–30.

34. Brenner, "Agrarian Roots of European Capitalism," p. 17.

35. Ruben, "Marxism and Dialectics," in John Mepham and D.-H. Ruben, eds., *Issues in Marxist Philosophy*, vol. 1 (Atlantic Highlands, N.J.: Humanities, 1979), pp. 50–51.

36. Laird Addis notes this parallel in *The Logic of Society* (Minneapolis: University of Minnesota Press, 1975), chapter 11.

37. See Steven Lukes, "Methodological Individualism Reconsidered," in Alan Ryan, ed., *The Philosophy of Social Explanation* (Oxford: Oxford University Press, 1973) for a survey of some of the literature on this issue.

38. For a dissenting view, however, see Richard Miller, "Methodological Individualism and Social Explanation," *Philosophy of Science* 45 (1978):387–414.

39. D.-H. Ruben, "Marx, Necessity, and Science," in G. H. R. Parkinson, ed., *Marx and Marxisms* (Cambridge: Cambridge University Press, 1982), p. 40.

40. Ruben, "Marx, Necessity, and Science," pp. 42–48.

41. Roemer, "Methodological Individualism and Deductive Marxism," p. 513.

42. A definitive description of the H-D model is found in Carl Hempel's "Scientific Explanation," in his *Aspects of Scientific Explanation* (New York: Free Press, 1965). It must be noted that Hempel's account is much more qualified than this summary statement indicates.

43. Michael Scriven emphasizes the pragmatic features of explanation in "Explanations, Predictions, and Laws" in Herbert Feigl and Grover Maxwell, eds. *Minnesota Studies in the Philosophy of Science*, volume 3 (Minneapolis: University of Minnesota Press, 1962). See especially pp. 176–90 for his criticisms of the H-D model. "What is a scientific explanation? It is a topically unified communication, the content of which imparts understanding of some scientific phenomenon" (p. 224).

Chapter 6. Evidence and Justification

1. For a good account of changing positions on theory and justification, see Frederick Suppe, Introduction, in Suppe, ed., *The Structure of Scientific Theories* (Urbana: University of Illinois Press, 1974).

2. The hypothetico-deductive model is admirably described and developed in Carl Hempel's *Aspects of Scientific Explanation* (New York: Free Press, 1965), especially in "The Function of General Laws in History" and "Confirmation, Induction, and Rational Belief." Clark Glymour's *Theory and Evidence* (Princeton: Princeton University Press, 1980) offers sharp criticisms of the H-D model from the standpoint of enlightened logical positivism (pp. 10–63).

3. Consider, though, Paul Feyerabend's radical downplaying of the importance of empirical evidence and scientific method in science in *Against Method* (London: New Left Books, 1975), pp. 17–68.

4. W. H. Newton-Smith's *The Rationality of Science* (Boston: Routledge & Kegan Paul, 1981) discusses second-order criteria of theory choice, including fertility, intertheory support, smoothness, internal consistency, compatibility with well-grounded metaphysical beliefs, and simplicity (pp. 227–32).

5. Thomas Kuhn, *Structure of Scientific Revolutions*, 2nd ed. (Chicago: University of Chicago Press, 1970); Norwood Hanson, *Patterns of Discovery: An Inquiry into the Conceptual Foundations*

of Science (Cambridge: Cambridge University Press, 1965); and Paul Feyerabend, *Against Method*.

6. Newton-Smith's *The Rationality of Science* is a vigorous criticism of Kuhn, Laudan, Lakatos, Feyerabend, and "noncognitivist" theories of science generally. This book is important for its defense of the idea that scientific theories may be justified on rational grounds. See also David Thomas's consideration of some of the ways in which background assumptions intrude on observation in social science in *Naturalism and Social Science* (Cambridge: Cambridge University Press, 1979), chapters 2 and 3.

7. J. S. Mill, *A System of Logic*, books III:XI and VI:IX.

8. See Daniel Hausman's useful discussion of Mill's deductive method in "John Stuart Mill's Philosophy of Economics," *Philosophy of Science* 48 (1981):363–85.

9. Samuelson's view is that Marx's economic reasoning is not worthy of serious consideration: "It used to be thought that Marx's great achievement was the more-than-a-million-word *Das Kapital*, which scientifically foretold the *laws of motion of capitalism* unto its inevitable end. . . . Such is the vulgar view of what Marxism was all about." "But remember that Marx was a 19th century German philosopher." "Careful analytical audit of the logic of Karl Marx's system will show that he was not merely unlucky in his predicted factual laws of motion of capitalist development. His purported laws . . . are *not cogently derivable* from Marx's own conceptual schemes." Paul Samuelson, *Economics*, 11th ed. (New York: McGraw-Hill, 1980), pp. 795, 800, 806.

10. Ernest Mandel offers sharp criticism of Samuelson's understanding of Marx; Introduction, *Capital II*, p. 23.

11. Wassily Leontief, "The Significance of Marxian Economics for Present-Day Economic Theory," in David Horowitz, ed., *Marx and Modern Economics* (New York: Modern Reader, 1968), p. 98.

12. Schumpeter's assessment is advanced throughout his *History of Economic Analysis* (Oxford: Oxford University Press, 1954), for example at pp. 439, 441, 877–79.

13. Michio Morishima, *Marx's Economics* (Cambridge: Cambridge University Press, 1973), pp. 2–5.

14. Ernest Mandel, Introduction, *Capital I*, p. 23. See also Mandel's discussion of the empirical credentials of Marx's system in *Late Capitalism* (London: Verso, 1975), pp. 13–42.

15. Chapter 7 contains an extensive discussion of Popper's charges of unfalsifiability.

16. Roy Bhaskar, "On the Possibility of Social Scientific Knowledge and the Limits of Naturalism," in John Mepham and D.-H. Ruben, eds., *Issues in Marxist Philosophy*, vol. 3 (Atlantic Highlands, N.J.: Humanities, 1979), p. 128.

17. Derek Sayer, *Marx's Method* (Atlantic Highlands, N.J.: Humanities, 1979), p. 139.

18. Allen Wood, *Karl Marx* (London: Routledge & Kegan Paul, 1981), p. 228.

19. Richard Hudelson, "Marx's Empiricism," *Philosophy of the Social Sciences* 12 (1982):242.

20. Ibid.

21. See, however, James Farr's critique of Hudelson's position in "Marx No Empiricist," *Philosophy of the Social Sciences* 13 (1983):465–72. For reasons presented in this chapter and in chapter 4 I am in closer agreement with Hudelson rather than Farr.

22. Mandel, *Late Capitalism*, p. 15.

23. Sayer, *Marx's Method*, p. 81.

Chapter 7. Falsifiability and Idealism

1. The falsifiability thesis is developed throughout Popper's works. For a simple statement see *Conjectures and Refutations: The Growth of Scientific Knowledge* (New York: Basic, 1965), pp. 36–37, and *The Logic of Scientific Discovery* (New York: Harper, 1968), pp. 78–92. For Popper's application of falsifiability to Marx, see *Conjectures and Refutations*, p. 37, and *The Open Society and Its Enemies*, vol. 2 (Princeton: Princeton University Press, 1971), pp. 187–89.

2. Karl Popper, *Conjectures and Refutations*, p. 36.

3. Ibid., p. 32.

4. Karl Popper, *Open Society and Its Enemies*, vol. 2. pp. 105, 109, 144.

5. Indeed, Popper seems to recognize this point when he writes that Marx's argument for the inevitability of socialism involves three stages, and that "*Capital* elaborates only . . . the 'first step' of this argument." *The Open Society and Its Enemies*, p. 136.

6. Ibid., p. 187.

7. Ernest Mandel summarizes some twentieth-century data on the behavior of the rate of profit in *Marxist Economic Theory*, vol. 1 (New York: Monthly Review, 1970), pp. 166-170.

8. Carl Hempel, *Philosophy of Natural Science* (Englewood Cliffs, N.J.: Prentice-Hall, 1966), pp.28-30.

9. Ibid., p. 30.

10. Popper, *Conjectures and Refutations*, p. 288.

11. Larry Laudan's *Progress and Its Problems* (Berkeley: University of California Press, 1977) provides a good critical summary of the views of philosophers of science who give pride of place to the idea of a research program or paradigm.

12. "The initial stage, the act of conceiving or inventing a theory, seems to me neither to call for logical analysis nor to be susceptible to it. The question how it happens that a new idea occurs to a man . . . may be of great interest to empirical psychology; but it is irrelevant to the logical analysis of scientific knowledge. . . . Accordingly I shall distinguish sharply between the process of conceiving a new idea, and the methods and results of examining it logically." Popper, *The Logic of Scientific Discovery*, p. 31.

13. See Imre Lakatos, "Falsification and the Methodology of Scientific Research Programmes," in Lakatos and Musgrave, eds., *Criticism and the Growth of Knowledge* (Cambridge: Cambridge University Press, 1970), pp. 91-195, for a definitive statement of the notion of a research program. Larry Laudan's *Progress and Its Problems* also discusses and extends the concept of a research program (pp. 70-120).

14. Lakatos, "Falsification and Methodology," pp. 132-38. Although I believe that Lakatos's criterion of progressiveness shows Marx's social theory to be genuinely scientific, it must be noted that Lakatos does not ("Methodology," p. 92 and elsewhere).

15. Ibid., pp. 116-22.

16. Ernest Mandel marshalls some of the empirical support available for Marx's economic analysis in his introduction to *Capital II*, pp. 22-24 and 80-86. His treatment is of special interest because it directly responds to Popper's charge of unfalsifiability.

17. Consider Michio Morishima, *Marx's Economics* (Cambridge: Cambridge University Press, 1973); J. H. Westergaard, *Class in A Capitalist Society* (New York: Basic, 1976); James O'Connor, *The Fiscal Crisis of the State* (New York: St. Martin's, 1973), and Samuel Bowles and Herbert Gintis, *Schooling in Capitalist America* (New York: Basic, 1976).

18. E. P. Thompson, *The Poverty of Theory and Other Essays* (New York: Monthly Review, 1978) (referred to hereafter as *PT*).

19. Thompson charges that Althusser attempts to reproduce the reality of the social system on the basis of a lofty "theory" without the trouble of historical or empirical investigation. This effort represents an insidious form of Marxist idealism rather than empirical social science in that it holds theory sovereign over empirical inquiry (*PT*, pp. 4-5, 13, 188, etc.).

20. Perry Anderson's *Arguments within English Marxism* (London: New Left Books, 1980) discusses the justice of Thompson's charges against Althusser at some length.

21. Anderson (*Arguments*, pp. 59-64) rejects this assessment of Marx's work, holding (as I do) that Marx's economic analysis in *Capital* is a reasonable scientific task, given the premises of historical materialism.

22. See Nicos Poulantzas, *Political Power and Social Classes*, trans. Timothy O'Hagan (Lon-

don: New Left Books, 1975), pp. 11-98, for some discussion of the notion of "regional theories" in application to political theory.

23. Barry Hindess and Paul Hirst, *Pre-Capitalist Modes of Production* (London: Routledge & Kegan Paul, 1975).

24. David Hull's *Philosophy of Biological Science* (Englewood Cliffs, N.J.: Prentice-Hall, 1974), chapter 2, offers a useful discussion of the relation between different levels of analysis in biology.

25. For a detailed discussion of the forces and relations of production in Marx's account of capitalism, see G. A. Cohen, *Karl Marx's Theory of History: A Defence* (Princeton: Princeton University Press, 1978), chapters 2 and 3.

26. For illustrations of this point, see Ralph Miliband, *The State in Capitalist Society* (New York: Basic, 1969), chapter 2, and *Capitalist Democracy in Britain* (Oxford: Oxford University Press, 1982), chapters 2 and 3.

27. See Miliband, *The State in Capitalist Society*, pp. 153-65, for examples of structural constraints on law and political policy.

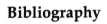

Bibliography

Bibliography

See Bibliographical Note on pp. xi-xii for references to Marx's writings.

Achinstein, Peter. *Concepts of Science*. Baltimore: Johns Hopkins University Press, 1968.
Addis, Laird. "The Individual and the Marxist Philosophy of History." In May Brodbeck, ed., *Readings in the Philosophy of the Social Sciences*. New York: Macmillan, 1968.
——. *The Logic of Society*. Minneapolis: University of Minnesota Press, 1975.
Althusser, Louis. *For Marx*. Translated by Ben Brewster. London: New Left Books, 1969.
Althusser, Louis, and Etienne Balibar. *Reading Capital*. Translated by Ben Brewster. London: New Left Books, 1970.
Anderson, Perry. *Lineages of the Absolutist State*. London: New Left Books, 1974.
——. *Passages from Antiquity to Feudalism*. London: New Left Books, 1974.
——. *Arguments within English Marxism*. London: New Left Books, 1980.
Arrow, Kenneth. *Social Choice and Individual Values*. New York: Wiley, 1963.
Atkinson, R. F. "Historical Materialism." In G. H. R. Parkinson, ed., *Marx and Marxisms*. Cambridge: Cambridge University Press, 1982.
Bahro, Rudolph. *The Alternative in Eastern Europe*. London: New Left Books, 1981.
Ball, Terence, and James Farr, eds. *After Marx*. Cambridge: Cambridge University Press, 1984.
Barry, Brian. *Sociologists, Economists, and Democracy*. Chicago: University of Chicago Press, 1978.
Berlin, Isaiah. *Karl Marx, His Life and Environment*. 3rd ed. London: Oxford University Press, 1963.
Bhaskar, Roy. *A Realist Theory of Science*. Atlantic Highlands, N.J.: Humanities, 1978.
——. "On the Possibility of Social Scientific Knowledge and the Limits of Naturalism." In John Mepham and D.H. Ruben, eds., *Issues in Marxist Philosophy*. Vol. 3. Atlantic Highlands, N.J.: Humanities, 1979.
Blackburn, Robin, ed. *Ideology in Social Science*. New York: Vintage, 1973.
Blaug, Mark. *Economic Theory in Retrospect*. Homewood, Ill.: Irwin, 1968.
Bloch, Marc. *French Rural History*. Berkeley: University of California Press, 1966.
——. *Land and Work in Medieval Europe*. London: Routledge & Kegan Paul, 1967.
Böhm-Bawerk, Eugen von. *Karl Marx and the Close of His System*. London: Merlin, 1975.
Bowles, Samuel, and Herbert Gintis. *Schooling in Capitalist America*. New York: Basic, 1976.
Braverman, Harry. *Labor and Monopoly Capital*. New York: Monthly Review, 1974.
Brenkert, George. *Marx's Ethics of Freedom*. London: Routledge & Kegan Paul, 1983.

Brenner, Robert. "Agrarian Class Structure and Economic Development in Pre-Industrial Europe." *Past and Present* 70 (1976):30–75.
——. "The Agrarian Roots of European Capitalism." *Past and Present* 97 (1982):16–113.
Brewer, Anthony. *A Guide to Marx's Capital.* Cambridge: Cambridge University Press, 1984.
Brodbeck, May, ed. *Readings in the Philosophy of the Social Sciences.* New York: Macmillan, 1968.
Brown, Harold I. *Perception, Commitment, and Theory.* Chicago: University of Chicago Press, 1979.
Buchanan, Allen. "Revolutionary Motivation and Rationality." *Philosophy and Public Affairs* 9 (1979):59–82.
——. *Marx and Justice.* Totowa, N.J.: Littlefield, Adams, 1982.
Burtt, E. A. *The Metaphysical Foundations of Modern Science.* 2nd ed. Garden City, N.Y.: Doubleday, 1954.
Carver, Terrell. *Marx's Social Theory.* Oxford: Oxford University Press, 1982.
——. "Marx, Engels and Scholarship." *Political Studies* 32 (1984):249–56.
Chomsky, Noam. Review of *Verbal Behavior* by B. F. Skinner. *Language* 35:1 (1959):26–58.
——. *Aspects of the Theory of Syntax.* Cambridge, Mass.: MIT Press, 1965.
Cohen, G. A. "On Some Criticisms of Historical Materialism." *Aristotelian Society Suppl. Volume* 44 (1970):121–41.
——. "Karl Marx and the Withering Away of Social Science." *Philosophy and Public Affairs* 1 (1972):182–203.
——. "Marx's Dialectic of Labor." *Philosophy and Public Affairs* 3 (1974):235–61.
——. *Karl Marx's Theory of History: A Defence.* Princeton: Princeton University Press, 1978.
——. "Capitalism, Freedom and the Proletariat." In Alan Ryan, ed., *The Idea of Freedom.* Oxford: Oxford University Press, 1979.
——. "The Labor Theory of Value and the Concept of Exploitation." *Philosophy and Public Affairs* 8 (1979):338–60.
——. "More on Exploitation and the Labour Theory of Value." *Inquiry* 26 (1983):309–31.
——. "Functional Explanation: Reply to Elster." *Political Studies* 28 (1980):129–35.
——. "Functional Explanation, Consequence Explanation, and Marxism." *Inquiry* 25 (1982):27–56.
——. Reply to Elster on "Marxism, Functionalism, and Game Theory." *Theory and Society* 11 (1982):483–95.
Cohen, Joshua. Review of *Karl Marx's Theory of History* by G. A. Cohen. *The Journal of Philosophy* 79 (1982):253–73.
Cohen, Marshall, Thomas Nagel, and Thomas Scanlan, eds. *Marx, Justice, and History.* Princeton: Princeton University Press, 1980.
Croce, Benedetto. *Essays on Marx and Russia.* New York: Frederick Ungar, 1966.
Cutler, Antony, et al. *Marx's Capital and Capitalism Today.* Vols. 1, 2. London: Routledge & Kegan Paul, 1977, 1978.
Danto, Arthur, and Sidney Morgenbesser, eds. *Philosophy of Science.* New York: World Publishing, 1960.
Dawkins, Richard. *The Selfish Gene.* Oxford: Oxford University Press, 1976.
Elster, Jon. *Logic and Society.* Chichester: Wiley, 1978.
——. *Ulysses and the Sirens.* Cambridge: Cambridge University Press, 1979.
——. "Cohen on Marx's Theory of History." *Political Studies* 28:1 (1980):121–28.
——. "Belief, Bias, and Ideology." In Martin Hollis and Steven Lukes, eds., *Rationality and Relativism.* Cambridge, Mass.: MIT Press, 1982.
——. "Marxism, Functionalism, and Game Theory." *Theory and Society* 11 (1982):453–82.
——. "A Paradigm for the Social Sciences?" *Inquiry* 25 (1982):378–85.

——. "Roemer versus Roemer: A Comment on 'New Directions in the Marxian Theory of Exploitation.' " *Politics and Society* 11 (1982):363–74.

——. "Sour Grapes – Utilitarianism and the Genesis of Wants." In A. K. Sen and Bernard Williams, eds., *Utilitarianism and Beyond*. Cambridge: Cambridge University Press, 1982.

——. *Explaining Technical Change*. Cambridge: Cambridge University Press, 1983.

——. "Historical Materialism and Economic Backwardness." In Terence Ball and James Farr, eds., *After Marx*. Cambridge: Cambridge University Press, 1984.

Farr, James. "Marx No Empiricist." *Philosophy of the Social Sciences* 13 (1983):465–72.

Feigl, Herbert, and Grover Maxwell, eds. *Minnesota Studies in the Philosophy of Science*. Vol. 3. Minneapolis: University of Minnesota Press, 1962.

Feyerabend, Paul. *Against Method*. London: New Left Books, 1975.

Fisk, Milton. "Dialectic and Ontology." In John Mepham and D.-H. Ruben, eds., *Issues in Marxist Philosophy*. Vol. 3. Atlantic Highlands, N.J.: Humanities, 1979.

Geras, Norman. "Essence and Appearance: Aspects of Fetishism in Marx's Capital." *New Left Review* 65 (1971):69–85.

——. "Marx and the Critique of Political Economy." In Robin Blackburn, ed., *Ideology in Social Science*. New York: Vintage, 1973.

Glymour, Clark. *Theory and Evidence*. Princeton: Princeton University Press, 1980.

Godelier, Maurice. *Rationality and Irrationality in Economics*. Translated by Brian Pearce. New York: Monthly Review, 1978.

Gottheil, Fred. *Marx's Economic Predictions*. Evanston, Ill,: Northwestern University Press, 1966.

Gramsci, Antonio. *Selections from the Prison Notebooks of Antonio Gramsci*. Translated and edited by Q. Hoare and G. Nowell-Smith. New York: International Publishers, 1971.

Hahn, Frank, and Martin Hollis, eds. *Philosophy and Economic Theory*. Oxford: Oxford University Press, 1979.

Hanson, Norwood. *Patterns of Discovery: An Inquiry into the Conceptual Foundations of Science*. Cambridge: Cambridge University Press, 1965.

Hausman, Daniel. "John Stuart Mill's Philosophy of Economics." *Philosophy of Science* 48 (1981):363–85.

Hegel, G. W. F. *The Philosophy of Right*. Translated by T. M. Knox. Oxford: Oxford University Press, 1969.

——. *Reason in History*. Translated by Robert S. Hartman. Indianapolis: Bobbs Merrill, 1953.

——. *Hegel's Logic. Part One of the Encyclopaedia of the Philosophical Sciences (1830)*. Translated by William Wallace. Oxford: Oxford University Press, 1975.

——. *Lectures on the Philosophy of World History*. Translated by H. B. Nisbet. Cambridge: Cambridge University Press, 1975.

Hempel, Carl. *Aspects of Scientific Explanation*. New York: Free Press, 1965.

——. *Philosophy of Natural Science*. Englewood Cliffs, N.J.: Prentice-Hall, 1966.

Hindess, Barry, and Paul Hirst. *Pre-Capitalist Modes of Production*. London: Routledge & Kegan Paul, 1975.

Hirst, Paul. "The Necessity of Theory." *Economy and Society* 8:4 (1979):417–45.

Hobsbawn, Eric. Introduction to Karl Marx, *Pre-Capitalist Economic Formations*. New York: International Publishers, 1972.

——. "Karl Marx's Contribution to Historiography." In Robin Blackburn, ed., *Ideology in Social Science*. New York: Vintage, 1973.

Hodges, Donald. *The Bureaucratization of Socialism*. Amherst, Mass.: University of Massachusetts Press, 1981.

Hollis, Martin, and Steven Lukes, eds. *Rationality and Relativism*. Cambridge, Mass.: MIT Press, 1982.

Holmstrom, Nancy. "Marx and Cohen on Exploitation and the Labor Theory of Value." *Inquiry* 26 (1983):287-307.

Hook, Sidney. *From Hegel to Marx.* Ann Arbor: University of Michigan Press, 1962.

Horowitz, David, ed. *Marx and Modern Economics.* New York: Modern Reader, 1968.

Hudelson, Richard. "Marx's Empiricism." *Philosophy of the Social Sciences* 12 (1982):241-53.

Hull, David. *Philosophy of Biological Science.* Englewood Cliffs, N.J.: Prentice-Hall, 1974.

Hunt, E. K., and Jesse G. Schwartz. *A Critique of Economic Theory.* Baltimore: Penguin Books, 1972.

Keat, Russell, and John Urry. *Social Theory as Science.* London: Routledge & Kegan Paul, 1975.

Kuhn, Thomas. *The Structure of Scientific Revolutions.* 2nd ed. Chicago: University of Chicago Press, 1970.

Lakatos, Imre, and Alan Musgrave, eds. *Criticism and the Growth of Knowledge.* Cambridge: Cambridge University Press, 1970.

Laudan, Larry. *Progress and Its Problems.* Berkeley: University of California Press, 1977.

——. *Science and Hypothesis.* Dordrecht: D. Reidel, 1981.

Lenin, V. I. *The Lenin Anthology.* Edited by Robert Tucker. New York: Norton, 1975.

——. *Marx, Engels, Marxism.* Peking: Foreign Languages Press, 1978.

——. *On the Question of Dialectics.* Moscow: Progress Publishers, 1980.

Leontief, Wassily. "The Significance of Marxian Economics for Present Day Economic Theory." In David Horowitz, ed., *Marx and Modern Economics.* New York: Modern Reader, 1968.

Luce, R. D., and Howard Raiffa. *Games and Decisions.* New York: Wiley, 1958.

Lukács, Georg. *History and Class Consciousness.* Cambridge: MIT Press, 1971.

Lukes, Steven. "Methodological Individualism Reconsidered." In Alan Ryan, ed., *The Philosophy of Social Explanation.* Oxford: Oxford University Press, 1973.

Mackie, J. L. *The Cement of the Universe.* Oxford: Oxford University Press, 1974.

Mandel, Ernest. *Marxist Economic Theory.* Vol. 1. New York: Monthly Review, 1970.

——. *Late Capitalism.* London: Verso, 1978.

McMurtry, John. *The Structure of Marx's World-View.* Princeton: Princeton University Press, 1977.

Meek, Ronald. *Studies in the Labor Theory of Value.* 2nd ed. New York: Monthly Review, 1956.

Mepham, John, and D.-H. Ruben, eds. *Issues in Marxist Philosophy.* 4 vols. Atlantic Highlands, N.J.: Humanities, 1979.

Miliband, Ralph. *The State in Capitalist Society.* New York: Basic, 1969.

——. *Marxism and Politics.* Oxford: Oxford University Press, 1977.

——. *Capitalist Democracy in Britain.* Oxford: Oxford University Press, 1982.

Mill, John Stuart. *A System of Logic.*

Miller, Richard. "The Consistency of Historical Materialism." *Philosophy and Public Affairs* 4 (1975):390-409.

——. "Methodological Individualism and Social Explanation." *Philosophy of Science* 45 (1978): 387-414.

——. "Productive Forces and the Forces of Change." *Philosophical Review* 90 (1981):91-117.

——. *Analyzing Marx.* Princeton: Princeton University Press, 1984.

Moore, Stanley. "Marx and Lenin as Historical Materialists." *Philosophy and Public Affairs* 4 (1975):171-94.

——. "A Consistency Proof for Historical Materialism." *Philosophy and Public Affairs* 5 (1976):314-22.

Morishima, Michio. *Marx's Economics.* Cambridge: Cambridge University Press, 1973.

Nagel, Ernest. "The Meaning of Reduction in the Natural Sciences." In Arthur Danto and Sidney Morgenbesser, eds., *Philosophy of Science.* New York: World Publishing, 1960.

——. *The Structure of Science.* New York: Harcourt, Brace & World, 1961.

Nagel, Thomas. *The Possibility of Altruism.* Oxford: Oxford University Press, 1970.

Nell, Edward. "Economics – The Revival of Political Economy." In Robin Blackburn, ed., *Ideology in Social Science*. New York: Vintage, 1973.

Newton-Smith, W. H. *The Rationality of Science*. Boston: Routledge & Kegan Paul, 1981.

Nield, Keith, and John Seed. "Theoretical Poverty or the Poverty of Theory." *Economy and Society* 8 (1979):383–416.

Noble, James. "Marxian Functionalism." In Terence Ball and James Farr, eds., *After Marx*. Cambridge: Cambridge University Press, 1984.

O'Connor, James. *The Fiscal Crisis of the State*. New York: St. Martin's, 1973.

Ollman, Bertell. *Alienation: Marx's Conception of Man in Capitalist Society*. Cambridge: Cambridge University Press, 1972.

Olson, Mancur. *The Logic of Collective Action: Public Goods and the Theory of Groups*. Cambridge, Mass.: Harvard University Press, 1965.

Parfit, Derek. *Reasons and Persons*. Oxford: Oxford University Press, 1984.

Parkinson, G. H. R., ed. *Marx and Marxisms*. Cambridge: Cambridge University Press, 1982.

Popkin, Samuel. *The Rational Peasant*. Berkeley: University of California Press, 1979.

Popper, Karl. *The Poverty of Historicism*. London: Routledge & Kegan Paul, 1961.

——. *Conjectures and Refutations: The Growth of Scientific Knowledge*. New York: Basic, 1965.

——. *The Open Society and Its Enemies*. Vol. 2. Princeton: Princeton University Press, 1971.

——. *The Logic of Scientific Discovery*. New York: Harper, 1968.

Poulantzas, Nicos. *Political Power and Social Classes*. Translated by Timothy O'Hagan. London: New Left Books, 1975.

Rader, Melvin. *Marx's Interpretation of History*. Oxford: Oxford University Press, 1979.

Rapoport, Anatol. *Two-Person Game Theory*. Ann Arbor: University of Michigan Press, 1966.

Rapoport, Anatol, and A. M. Chammah. *Prisoners' Dilemma: A Study in Conflict and Cooperation*. Ann Arbor: University of Michigan Press, 1965.

Rawls, John. *A Theory of Justice*. Cambridge: Harvard University Press, 1971.

Reynolds, Terry. "Medieval Roots of the Industrial Revolutions." *Scientific American* 251 (1984):122–30.

Robinson, Joan. *An Essay in Marxian Economics*. New York: St. Martin's Press, 1942.

Roemer, John. *Analytical Foundations of Marxian Economic Theory*. Cambridge: Cambridge University Press, 1981.

——. "R. P. Wolff's Reinterpretation of Marx's Labor Theory of Value." *Philosophy and Public Affairs* 12 (1981):70–83.

——. *A General Theory of Exploitation and Class*. Cambridge, Mass.: Harvard University Press, 1982.

——. "Methodological Individualism and Deductive Marxism." *Theory and Society* 11 (1982): 513–20.

——. "New Directions in the Marxian Theory of Exploitation." *Politics and Society* 11 (1982): 253–87.

——. "Property Relations vs. Surplus Value in Marxian Exploitation." *Philosophy and Public Affairs* 11 (1982):281–313.

Rosdolsky, Roman. *The Making of Marx's Capital*. London: Pluto Press, 1977.

Rubel, Maximilien, and Margaret Manale. *Marx without Myth*. New York: Basic, 1975.

Ruben, David-Hillel. "Marxism and Dialectics." in John Mepham and D.-H. Ruben, eds., *Issues in Marxist Philosophy*. Vol. 1. Atlantic Highlands, N.J.: Humanities, 1979.

——. *Marxism and Materialism*. 2nd ed. Atlantic Highlands, N.J.: Humanities, 1979.

——. "Marx, Necessity, and Science." In G. H. R. Parkinson, ed., *Marx and Marxisms*. Cambridge: Cambridge University Press, 1982.

Rubin, I. I. *Essays on Marx's Theory of Value*. Translated by Milos Samardzija and Fredy Perlman. Originally published in 1928. Detroit: Black and Red, 1972.

Rudner, Richard. *Philosophy of Social Science*. Englewood Cliffs, N.J.: Prentice-Hall, 1966.
Ryan, Alan, ed. *The Philosophy of Social Explanation*. Oxford: Oxford University Press, 1973.
——. *The Idea of Freedom*. Oxford: Oxford University Press, 1979.
Samuelson, Paul. *Economics*. 11th ed. New York: McGraw-Hill, 1980.
Sayer, Derek. *Marx's Method*. Atlantic Highlands, N.J.: Humanities, 1979.
Scheffler, Israel. *The Anatomy of Inquiry*. Indianapolis: Bobbs-Merrill, 1963.
——. *Science and Subjectivity*. Indianapolis: Bobbs-Merrill, 1967.
Schelling, Thomas. *Micromotives and Macrobehavior*. New York: Norton, 1978.
Schumpeter, Joseph. *History of Economic Analysis*. Oxford: Oxford University Press, 1954.
Scott, James. *The Moral Economy of the Peasant*. New Haven: Yale University Press, 1976.
Scriven, Michael. "Explanations, Predictions, and Laws." In Herbert Feigl and Grover Maxwell, eds., *Minnesota Studies in the Philosophy of Science*. Vol. 3. Minneapolis: University of Minnesota Press, 1962.
Sen, A. K. *Collective Choice and Social Welfare*. San Francisco: Holden-Day, 1970.
——. "Rational Fools." In Frank Hahn and Martin Hollis, eds., *Philosophy and Economic Theory*. Oxford: Oxford University Press, 1979.
Sen, Amartya and Bernard Williams, eds. *Utilitarianism and Beyond*. Cambridge: Cambridge University Press, 1982.
Shaw, William. *Marx's Theory of History*. Stanford, Calif.: Stanford University Press, 1978.
——. "Marxism, Revolution and Rationality." In Terence Ball and James Farr, eds., *After Marx*. Cambridge: Cambridge University Press, 1984.
Simon, Herbert. *The Sciences of the Artificial*. 2nd ed. Cambridge: MIT Press, 1981.
Sraffa, Piero. *The Production of Commodities by Means of Commodities*. Cambridge: Cambridge University Press, 1960.
Steedman, Ian. *Marx after Sraffa*. London: New Left Books, 1977.
——, ed. *The Value Controversy*. London: New Left Books, 1981.
Suppe, Frederick, ed. *The Structure of Scientific Theories*. 2nd ed. Urbana: University of Illinois Press, 1977.
Sweezy, Paul. *The Theory of Capitalist Development*. New York: Monthly Review, 1968.
Szymanski, Albert. *The Capitalist State and the Politics of Class*. Cambridge, Mass.: Winthrop, 1978.
Thomas, David. *Naturalism and Social Science*. Cambridge: Cambridge University Press, 1979.
Thompson, E. P. *The Making of the English Working Class*. New York: Vintage, 1963.
——. *The Poverty of Theory and Other Essays*. New York: Monthly Review, 1978.
Tucker, Nancy Bernkopf. *Patterns in the Dust: Chinese-American Relations and the Recognition Controversy 1949–50*. New York: Columbia University Press, 1983.
Van Parijs, Philippe. *Evolutionary Explanation in the Social Sciences: An Emerging Paradigm*. Totowa, N.J.: Rowman & Littlefield, 1981.
——. "Functionalist Marxism Rehabilitated." *Theory and Society* 11 (1982):497–511.
——. "Why Marxist Economics Needs Microfoundations." *Review of Radical Political Economics* 156 (1983):111–24.
——. "Marxism's Central Puzzle." In Terence Ball and James Farr, eds., *After Marx*. Cambridge: Cambridge University Press, 1984.
Watkins, J. W. N. "Methodological Individualism and Social Tendencies." In May Brodbeck, ed., *Readings in the Philosophy of Social Sciences*. New York: Macmillan, 1968.
Weber, Max. *The Methodology of the Social Sciences*. Translated and edited by E. Shils and H. A. Finch. New York: Free Press, 1949.
——. *The Protestant Ethic and the Spirit of Capitalism*. New York: Scribner's, 1958.
Westergaard, J. H. *Class in a Capitalist Society*. New York: Basic, 1976.
Westfall, Richard. *The Construction of Modern Science*. New York: Wiley, 1971.

White, Lynn. *Medieval Technology and Social Change*. Oxford: Oxford University Press, 1962.

Wolff, Robert Paul. "A Critique and Reinterpretation of Marx's Labor Theory of Value." *Philosophy and Public Affairs* 10 (1981):89–120.

——. "Reply to Roemer." *Philosophy and Public Affairs* 12 (1982):84–88.

——. *Understanding Marx*. Princeton: Princeton University Press, 1984.

Wood, Allen. *Karl Marx*. London: Routledge & Kegan Paul, 1981.

Wright, Erik Olin. *Class, Crisis and the State*. London: Verso, 1978.

Young, Gary. "The Fundamental Contradiction of Capitalist Production." *Philosophy and Public Affairs* 5 (1976):196–234.

Index

Index

36, 62, 148, 170, 181, 198, 199, 211n35.
  See also Economic laws, historical spec-
  ificity of
Lenin, V. I., 11, 112, 114, 219n20
Leontief, Wassily, 78, 140, 166, 216n12
Lukács, Georg, 112, 196, 212n53, 217n3,
  219n21

Mackie, J. L., 27, 211n36
Macroexplanation, 127, 131. *See also* Micro-
  foundations
Mandel, Ernest, 23, 112, 167, 176, 186, 203,
  207n1, 219n22, 220n34, 225n10, 226n16
Marcuse, Herbert, 4
McMurtry, John, 5, 50, 54, 60, 213n11,
  214n22
Meek, Ronald, 216n6
Methodological writings, Marx's, 7, 92,
  199–200
Methodological individualism, 131, 149–52,
  223n19
Microfoundations, 6, 35, 57, 127–31,
  133–34, 139, 141, 151, 198
Microprocesses, 131
Miliband, Ralph, 37, 59, 186, 203, 212n50,
  218n13, 221n2, 227n26
Mill, John Stuart, 8, 25, 166, 196, 223n20
Miller, Richard, 49, 213n9, 224n38
Mode of production, 46–47, 163, 213n9
Moore, Stanley, 207n4
Morishima, Michio, 6, 78, 83, 167, 186,
  209n22

Nagel, Ernest, 14, 15
Nagel, Thomas, 222n17
Naturalism, 11, 13–16, 39, 170, 200, 209n6,
  211n44
Nell, Edward, 20
Newton, Isaac, 14, 15, 208n16
Newton-Smith, W. H., 224n4, 225n6
Noble, James, 129

O'Connor, James, 203, 223n21
Ollman, Bertell, 218n16
Olson, Mancur, 223n30

Pareto, Vilfredo, 196
Parfit, Derek, 222n17

Philosophical method, Marx's criticisms of,
  114–18, 120, 176, 196
Philosophy of science: 4–5; enlightened positivism, 14, 15, 183, 209n13; positivist,
  12, 14, 15–16, 155; postpositivist, 12, 16,
  156, 183
Political economy: orthodox, 94, 98, 127;
  vulgar, 108, 123, 218n7
Popkin, Samuel, 212n4, 223n18
Popper, Karl, 4, 5, 9, 11, 167–68, 177–87,
  199, 223n20, 225n1, 226n5
Poulantzas, Nicos, 7, 59, 186, 213n10,
  226n22
Predictions; scientific, 3, 10, 11, 12, 14, 16;
  role in *Capital*, 3, 9, 15, 158, 167–72,
  180, 199, 207n2
Prisoners' dilemma, 222n15
Productivity, 51, 132–34
Profit: 32, 69, 74–76, 143–44; falling rate of,
  3, 21, 26–27, 88, 135–39, 143–44, 169,
  180–81, 197, 223n25
Property relations, 30–31, 34, 42, 70, 91,
  134, 194, 202, 211n39. *See also* Relations
  of production
Proudhon, Pierre-Joseph, 8, 117, 215n44

Quesnay, François, 140

Rader, Melvin, 212n1
Rational-choice theory, 23, 129, 141, 145–46,
  198, 222n15, 223n28
Rationality, Marx's explanatory use of, 28,
  36, 121, 133, 139, 141–42, 144–45, 146,
  148, 223n19
Realism, scientific, 16, 20, 23–24, 210n26
Regularities. *See* Laws
Relations of production, 30–32, 42–46, 61,
  68, 99, 134, 196, 197. *See also* Economic
  structure
Relative immiseration, 137
Reproduction: schemes of, 6, 76–78, 139–41;
  simple, 77, 140, 146
Research program: 184–87, 203, 212n53;
  historical materialism as, 40, 60, 65–67,
  122
Revolution, 4, 179
Revolutionary motivation, 130, 214n20
Reynolds, Terry, 224n33

Daniel Little earned his Ph.D. in philosophy at Harvard University in 1977; he is currently associate professor of philosophy at Colgate University. During the 1985–86 academic year he served as visiting associate professor of philosophy at Wellesley College.

www.ingramcontent.com/pod-product-compliance
Ingram Content Group UK Ltd.
Pitfield, Milton Keynes, MK11 3LW, UK
UKHW021819110225
454967UK00009B/532